Rational Structural Design of Highway/Airport Pavements

Dawn of a new era

Rational Structural Design of Highway/Airport Pavements

New EVAPAVE, the Strongest & Toughest Paving Material

DINDIAL RAMSAMOOJ

authorHOUSE®

AuthorHouse™ LLC
1663 Liberty Drive
Bloomington, IN 47403
www.authorhouse.com
Phone: 1-800-839-8640

Published by AuthorHouse 07/26/2014

ISBN: 978-1-4918-5009-1 (sc)
ISBN: 978-1-4918-5008-4 (e)

Library of Congress Control Number: 2014902084

CONTENTS

Abstract

Rational design theories for highway and airport pavements are presented together with an invention of a vastly superior paving material comprising recycled Ethylene Vinyl Acetate (EVA) mixed and compacted with graded aggregates. The new material, called EVAPAVE, is five times stronger and tougher than asphalt concrete, and twice as strong and tough as high quality concrete. Fracture mechanics is used for determining the fatigue life, while the stress-dilatancy theory is used for the rutting of the pavement. The theories are then combined to obtain the interaction of fatigue and rutting. Several examples are presented to illustrate the design methodology. The new pavement will not have joints and will be the smoothest riding pavement, with huge savings in construction and maintenance and in vehicular fuel and maintenance costs, estimated to exceed $10 **billion** per year in the U.S. alone. It will outlast any other pavement by over seven times.

KEYWORDS: Recycled EVA; rational design; no joints; no ruts; economical.

Salutations unto thee O Lord

Without thee mind and speech will be naught

Salutations unto thee

To: Sachi, my wife, Patsy, my sister and other family members

PREFACE

"For he who innovates will have for his enemies all those who are well off under the existing order of things, and only lukewarm supporters in those who might be better off under the new." Niccolio Machiavelli, The Prince, 1910

*O*ver the centuries of civilization, there have been major improvements in the design and construction of roads and highways. The Romans were the greatest road builders in history. Over the past two centuries, there has been significant progress in highway pavement engineering. In the 1800's the macadam pavement was introduced in Britain. In 1908 the first concrete pavements (about 20 cm thick) were constructed in Detroit, and in 1924 the first asphalt concrete pavements (about 45 cm thick) was built in the United States of America. Technological advances in highway equipment and materials have made significant improvement in highway engineering.

Presently, all pavements are built using empirical or trial-and-error methods. Unfortunately, empirical methods inhibit real progress by obscuring the true parameters involved. Indeed, the most regrettable aspect of the empirical methods used, is that even the test data upon which they were invented have been found to have no relationship with theoretical parameters, thus rendering the task of innovative and lasting progress more difficult. This book will present new theories of pavement design and an innovative new paving material with vastly superior properties that will make pavements smoother riding, crack-free, more economical to construct, less costly for vehicular maintenance and fuel consumption, more enduring by requiring little or no maintenance, and enhance the environment by using recycled materials.

The ability to innovate lies in the mastery of the fundamentals. These fundamentals include foundation engineering, fracture mechanics, viscoelasticity, viscoplasticity and polymer engineering. Pavement design using new materials and new technology is the motivation behind this book. The purpose of this book is to introduce a systematic methodology for analysis and design of highway and airport pavements. Accordingly this book is intended for advanced undergraduate and graduate students, practicing engineers and professionals involved in the design and construction of such pavements.

The motivation behind this book stems from passion to improve the state-of-the-art from empirical to rational. This led to a strong desire to find new materials new technology and to design more efficiently and economically. Students should admire the power and the beauty of the analytical design equations over the empirical ones.

A major strength of this book is that the material is arranged in a pedagogically sound order. It gradually introduces the design process starting from the testing and preparation of the subgrade to the concepts of fracture mechanics, viscoelasticity/plasticity, permanent deformation, fatigue, rutting, and the interaction of fatigue and rutting. It then presents thermal stresses and the spacing of cracks which are deployed in the design process. Finally reliability analysis is introduced into the design process to account for the effects of statistical deviation of each design parameter.

The text gradually introduces new concepts and gives examples that use the **COMPUTER PROGRAMS** that are an integral part of the book, to help the reader in mastery of the new concepts as they are introduced. The exposition is clear, concise and written in simple language. All the computer programs have been actually run and tested numerous times over the years. I hope this book will make the design of highway and airport pavements exciting and rewarding.

The author's qualifications are:

D. V. Ramsamooj

B.Sc.(Eng.)Lond.,M.Sc.,Ph.D.,M.IC.E.,P.E.

Emeritus Professor, Cal. State University, Fullerton

1.

FUNDAMENTALS OF PAVEMENT ENGINEERING

"I believe that quality level is determined primarily by the actual design of the product itself, not by quality control in the production process."

Hideo Sigiura, Chairperson, Honda Motor Co.

1.1 Introduction

Rational design of highway and airport pavements is the application of *theory to practice* of the principles of fracture mechanics, elasticity and viscoelastic plasticity to engineering problems dealing with fatigue cracking and permanent deformation under repetitive highway loading, temperature changes and other environmental conditions in pavements. *Unification of the fatigue behavior of metals, concrete, asphalt concrete and plastics and the development of materials of greater strength and toughness are the guiding lights in this book.* It focuses on the development of sound theoretical principles against the impressive background of a vast amount of empirical data acquired over the past two centuries. The rapid advance of our knowledge of plastics (80,000) in recent years and in the use of computers has facilitated the development of theoretical applications to real highways. Awareness of the uncertainties of the fundamental assumptions used in computations, enables the engineer to anticipate the differences between reality and his original concept and assists him in making adaptations to the attain simple and realistic working knowledge. Simplicity is highly desirable in pavement design and generalization is the essence of simplicity. Yet unwarranted generalizations of non-ideal materials beyond the range of validity are avoided. Rigorous mathematical solutions published in the archival literature have often proved to be too complicated for realistic applications to highway problems so that they are often ignored.

The hallmark of a sound theory is that each parameter has a clear physical meaning and can be verified by credible existing published data and by realistic laboratory experiments that simulate real highway loading. As always definitions are most important, especially when new terms are being introduced. Accordingly *all definitions are expressed with clarity and are typed in italics.* Often published data are not comprehensive enough and the parameters used cannot be related to any theoretical parameters. The agreement of the theories with known *geometrical, environmental and boundary conditions* is essential. In this regard, dimensionless parameters are used whenever possible and are emphasized by **writing them in bold**. This facilitates the application of laboratory experiments, small-scale or full-scale experiments to real pavements with vehicular loading under the actual environmental conditions.

Every empirical rule expresses a probability and not a certainty. Yet these empirical rules offer some guide to the truth if used with good knowledge of statistics and the range of representation within the spectrum of applicability of highway materials.

1.2 Foundations of highways

Planning of a new highway route considers the geological, geotechnical and foundation requirements together with the environmental factors and compatibility of the foundation of the new pavement. Uniformity of the subgrade over a depth of about six feet is essential. If it is not, and there is no feasible alternative route, then the non-uniformities should be removed. For example , if proof rolling of the subgrade shows that there are relatively soft deposits of clay or organic matter, they should be dug out by a front-end rubber tired tractor or other suitable equipment and be replaced by easily compacted material such as sand or gravel. Loose sands need to be compacted to meet the usual specifications. After replacement or compaction, the subgrade should again be proof-rolled by smaller contact area rubber tired equipment to reveal the smaller non-uniformities. There are more modern imaging techniques and lasers that can do a better job. The bearing capacity, the total and differential settlement must be considered in choosing the type of highway pavement, flexible, rigid, or semi-rigid.

1.3 Pavement Type—Rigid, Flexible or Semi-rigid

Asphalt concrete, or bituminous stabilized sand or gravel are flexible pavements because of their low flexural modulus of elasticity (100-600) ksi (6890-4137 kPa), while concrete with a flexural modulus of 1,400 ksi (9,653 kPa) is rigid, so that a material with a modulus of 800,000-2,000,000 psi is semi-rigid. Flexible pavements are ideally suitable over hard subgrades, such as gravelly materials, while rigid pavements are ideally suitable for soft but *uniform* clay subgrades. A semi-rigid pavement such as a mixture of asphalt concrete and sulfur is suitable for subgrades of medium compressibility. A flexible pavement usually uses several layers to reduce the stress on the subgrade and rely on the strength and durability of the asphalt concrete surface as a wearing course. The basic philosophy of a flexible pavement is to utilize the inherent strength of the lower layers. A rigid pavement utilizes the high modulus of elasticity to spread the load over a larger area, thus reducing the stresses on the subgrade. A new paving material EVAPAVE will be introduced has a compressive modulus of 2 million psi and is classified as semi-rigid.

1.4 Seepage, Drainage and Flownets

Good drainage is an essential consideration in design. The permeability of asphalt concrete pavements in practice ranges from 0.01-10 ft. /day $(10^{-4}-1)$ (cm/s). All paving materials should be tested in the laboratory for permeability. Typical permeability of highway materials are given in Table 1.1

Table 1.1. Coefficients of saturated permeability of typical soils and highway materials

Soils (Coduto, 1999)	Saturated permeability, cm/s
Clean gravel	1-100
Sand-gravel mixtures	10^{-2}-10
Clean coarse sand	$10^{-2} - 1$
Fine sand	$10^{-3} - 10^{-1}$
Silty sand	$10^{-3} - 10^{-2}$
Clayey sand	$10^{-4} - 10^{-2}$
Silt	$10^{-6} - 10^{-3}$
Clay	$10^{-10} - 10^{-6}$

Table 1 .1 (cont'd)

Highway Materials (Carter and Bentley, 1991)	Saturated permeability, cm/s
Uniformly graded coarse aggregate	$0.4 - 40$
Well-graded aggregate without fines	$4(10^{-3}) - 4(10^{-1})$
Concrete sand, low dust content	$7(10^{-4}) - 7(10^{-2})$
Concrete sand, high dust content	$7(10^{-6}) - 10^{-4})$
Compacted silt	$7(10^{-8}) - 10^{-6})$
Compacted clay	$<10^{-7}$
Bituminous concrete	$4(10^{-6}) - 4(10^{-3})$
Portland cement concrete	$< 10^{-8}$

The surface layer of an AC pavement is not impermeable. The permeability of asphalt concrete k_{AC} varies from $10^{-6} - 10^{-4}$ (Table 1.1). Water entering the surface must be drained by the underlying layers (base and subbase) to prevent infiltration into the more vulnerable subgrade. Usually, it is the subbase that provides the most drainage. Assume that water seeps vertically through the surface and base courses. Neglecting the permeability of the subbase, the amount of water entering the subbase per unit area is

$$q_i = k_{AC} \, i \qquad\qquad (1.1)$$

in which i = the hydraulic gradient = 1.0. The total amount of water entering through a width b of the surface is

$$Q_i = k_{AC} b \qquad\qquad (1.2)$$

The drainage provided by the subbase from the seepage flownet in Figure 1.1 is:

$$Q_o = k_s \frac{N_f}{N_d} \Delta H \qquad\qquad (1.3)$$

where k_s = permeability of the subbase, and ΔH = the total head lost. From the flownet, Figure 1.1, $N_f / N_d = h / b$ and $\Delta H = h_s$. Therefore the outflow is

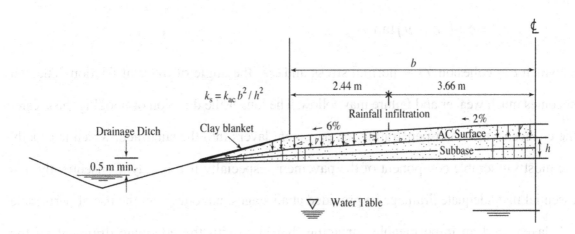

Figure 1.1 Seepage flownet for drainage highway pavement

$$Q_o = k_s \frac{h_s^2}{b} \qquad\qquad (1.4)$$

For a good design the outflow must be much greater than the inflow. Because of the very large differences in the permeabilities of highway materials due to the various grain sizes, a factor-of-safety $F_s = 10$ is recommended. Therefore the minimum permeability of the subbase, using a $F_s = 10$, is

$$k_s > 10 k_{AC} \frac{b^2}{h_s} \qquad (1.5)$$

When sufficient drainage is not provided, or the amount of cracks and ruts (permanent deformation) in the surface layer is too large, water may pond in the wheel paths

The most disastrous damage to pavements occurs when they become flooded, with the water forming a continuous film through the pores of the materials. It is conservatively assumed that the pressures from the vehicular tires are transmitted undiminished to the water, so that the pore water pressures may become as high as the tire pressures. Alternatively it may be roughly approximated using Boussinesq distribution with a Poisson's ratio of 0.495. The shear strength of a soil is given by

$$\tau_f = c + \left(\sigma - u \right) \tan\phi \qquad (1.6)$$

in which c = cohesion, σ = normal stress and ϕ = the angle of internal friction. The soil becomes much weaker and failure may follow. The longer the duration of flooding the greater the chance of the water penetrating the pavement layers into the subgrade, which is usually the most vulnerable component of the pavement, especially if it is clay. Consequently it is essential that adequate drainage be provided at all stages, not only with the use of permeable sub layers and an impermeable surfacing, but also with the adequate drainage by the provision of ditches that provide adequate drainage with the water level at least 0.5 m below the top of the subgrade, as shown in Figure 1.1.

1.5 Highways on expansive soil subgrades

Expansive soils are usually heavily overconsolidated clays that expand when the moisture content increase and shrink when the moisture content decrease similar to the rebound curve of a consolidation test. The swelling pressures are large, typically 0 – 96 kPa (0–2000 psf).

Highly expansive rocks can exert much higher pressures depending on the extent of the preconsolidation pressure. Soils or rocks containing montmorillonite are more expansive than kaolinite which are more expansive than illite. Highways constructed on expansive soils are subjected to large uplifting forces caused by the swelling. During the rainy season the moisture content of the subsoils increase and during the dry season the moisture content decreases. The zone of wetting and drying is called the active zone or the zone of desiccation, typically 0 to 5 m deep, as shown in Figure 1.2. The swelling potential is highest at the surface and decreases sharply with depth.

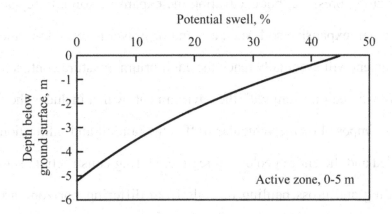

Figure 1.2 Potential swelling as a function of the depth

Civil engineers in general are stress-oriented professionals. As such they have difficulty with displacement-type phenomena. Because of the enormous pressure exerted by swelling soils, strengthening the superstructure or the upper layers of the highway is not an efficient way of dealing with the problem. The *swelling pressure is defined as that pressure at which increasing the moisture content does not cause further swelling.* In other words, the soil will not swell if the pressure exerted on it is equal or greater than the swelling pressure, as shown in Figure 1.3. Having explored the nature of the expansive soils, what is the highway engineer supposed to do?

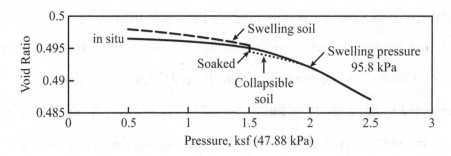

Figure 1.3 Illustration of the definition of swelling pressure

Clearly it is not feasible in a pavement to increase the downward pressure by as much as 100 kPa, the swelling pressure, but excavating the expansive soil and replacing sufficient thickness with a non-expansive soil, such as sand or gravel is feasible in most cases. The final moisture content will eventually reach the equilibrium moisture content, at which point the soil moisture is in equilibrium with the environment, which includes the stresses in the ground and those imposed by the vehicular traffic, the rainfall and infiltration of water, the drainage provided and the entire geological regime. During construction, selective grading including blending and cross hauling of soils from differing horizons are effective in reducing the swelling potential. Lime stabilization of the expansive subgrade, and increasing the strength and stiffness of the upper paving components are also used. An economic study should be performed before deciding on which option is most feasible.

Controlling the surface drainage is essential and relatively easy, but there could be rivers of flow beneath the surface depending on the subsurface geology and the manner of construction of the highway foundation. One common error of grading contractors is to excavate insufficiently deep to remove all of the highly compressible and weak or organic soils and replace them with compacted soils. When water flows in a certain drainage path, the fines are washed out or eroded, leaving the coarser material in the soils. As seen from Table 1.1, the permeability or hydraulic conductivity of a soil varies enormously with the particle size. For example, coarse sand has a permeability of up to 1 million times that of silt.

Example 1.1

A highway embankment is planned for construction over an existing subgrade. The soil is expansive and the potential swell as a function of the depth below the ground surface is shown in Fig.1.3. Calculate the total potential vertical rise and the probable differential swell, assuming that the differential swell is approximately one-half the total swell. What recommendations would you make for good performance of the future pavement?

Solution

The potential vertical rise caused by infiltration of water is equal to the area under the horizontal axis and the curve shown in Figure 1.2. The area is approximately 0.1 m (4 in.), so that the potential swelling is high. Excavate and remove the top 1 m would reduce the swell by the area under the curve up to a depth of 1 m, which is equal to 0.04 m, so that the potential vertical rise = 0.06 m (1.83 in). The excavated material is then replaced by a non-swelling material such well compacted as gravel or sand. The differential swell would then be 0.91 in. < 2.54 cm (1.0 in), which is acceptable. ← *Ans.*

1.6 Soil Suction

Water can be held in a soil above the water table by surface tension. Typical capillary suction varies with the type of soil as shown in Figure 1.4. Trees in the desert have been found to suck water by an exceedingly fine root system up to an astonishing 1500 cm above the water table. *The suction is defined as the height in cm above the water table to which water in a capillary tube inserted at a specific depth in the soil will rise.*

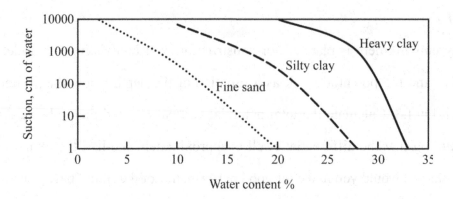

Figure 1.4 Suction as a function of water content

(drying condition), (Croney, D. & P., 1990)

Croney, D. and Croney, P (1990) found experimentally a relationship between the plasticity index and the compressibility factor that control the soil suction as shown in Figure 1.5. Eventually a soil reaches an equilibrium moisture content, which depends on the environment such as the drainage, rainfall, evaporation, and external stresses. *The soil suction is defined by*

$$s = u - \alpha\sigma_o \qquad (1.7)$$

where u = pore water pressure, σ_o = the overburden pressure, and α = the compressibility factor.

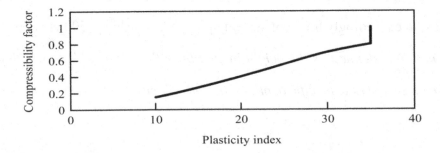

Figure 1.5 Relationship between compressibility and plasticity index (Croney, D. &P.)

Example 1.2

A subgrade consists of silty clay with density of 122 pcf, and plasticity index of 15. The water table is 6 ft. below the top of the subgrade as shown in Fig 1.6a. Determine the equilibrium water content of the subgrade at a depth of 3 ft. below the top of the subgrade. If the water table were to rise 3 ft. during a heavy rainy season, what would the water content be?

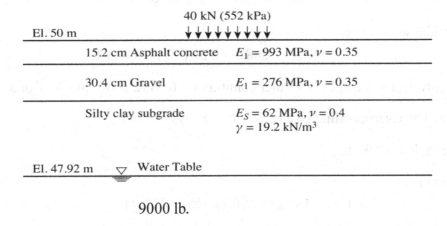

Figure 1.6a Typical flexible pavement

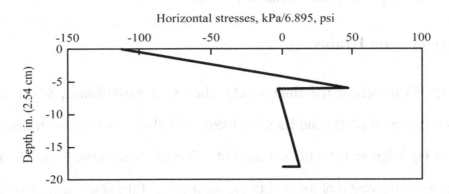

Figure 1.6b Stress distribution in the flexible pavement (*CHEVRON*)

Solution

(a) Water table at El. 94.0 ft.

From Fig. 2.5 the compressibility factor for a P. I = 15 is 0.27

The total stress at El.97 ft. is

$$\sigma_{ovb} = 144(0.83) + 3(122) = 455 \text{ psf}$$

The pore water pressure at El. 97 ft. is
$$u = -3(62.4) = -268.3 \text{ psf}$$

Therefore the soil suction
$$s = u - \alpha\sigma_{ovb} = -268.3 - 0.27(455) = -391 \text{ psf},$$

This is equivalent to a height of a water column of—6.26 ft. above the W. T or a suction

of 191 cm. The corresponding water content is 21.5% ← **Ans**

(b) Water table at El 97 ft.

The suction is
$$s = u - \alpha\sigma_{ovb} = 0 - 0.27(455) = -122.9 \text{ psf}$$

This is equivalent to a height of a water column of 1.97ft. above the W. T. or a suction of

60 cm. The corresponding water content is 24.0%. ← **Ans**

1.7 Theory of Sand Drains

Rendulic (1935) presented the fundamental theory of sand drains, which was later

summarized by Baron (1948) and Richart (1959). The plan of a sand drain installation is

presented in Fig 1.7(a) and (b). Let the radii of influence of the sand drain, the well, and

the distance from the center of the well to the outer edge of the smear zone ber$_e$, r_w and r_s

respectively. The consolidation theory (Terzaghi, 1943) for radial drainage is:

$$\frac{\partial u}{\partial t} = c_{vr}\left(\frac{\partial^2 u}{\partial t^2}\frac{1}{r}\frac{\partial u}{\partial r}\right) \tag{1.8}$$

where c_{vr} = the coefficient of consolidation in the radial direction, r = radial distance from the center of the drain and u = excess pore water pressure. It is assumed that drainage takes place only in the radial direction.

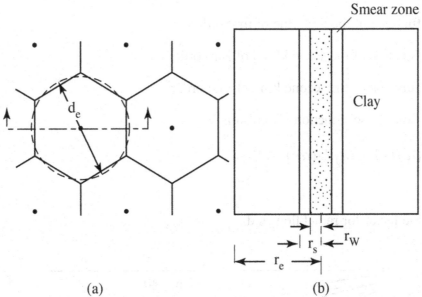

(a)   (b)

Figure 1.7a and 1.7b. Plan and cross-sectional elevation of sand drains

Eq. (1.8) is solved with the boundary conditions:

    (1) Time $t = 0$, $u = u_i$, the initial pore water pressure

    (2) Time $t > 0$, $u = 0$ at $r = r_w$.

    (3) Radius $r = r_e$, $\partial u / \partial r = 0$.

The excess pore water pressure at any time and radial distance is:

$$u = \sum_{\alpha_1,}^{\alpha = \infty} \frac{-2U_1(\alpha)U_0(\alpha r / r_w)}{\alpha\left\{n^2 U_0^2(\alpha n) - U_1^2(\alpha)\right\}} e^{-4\alpha^2 n^2 T_r} \qquad (1.9)$$

where $n = r_w$.

$$U_1(\alpha) = J_1(\alpha)Y_0(\alpha) - J_0(\alpha)Y_1(\alpha)$$
$$U_0(\alpha n) = J_0(\alpha n)Y_0(\alpha) - J_0(\alpha)Y_0(\alpha n)$$

$$U_0\left(\frac{\alpha r}{r_w}\right) = J_0\left(\frac{\alpha r}{r_w}\right)Y_0\left(\alpha\right) - J_0\left(\alpha\right)Y_0\left(\frac{\alpha r}{r_w}\right), where$$

J_0 = Bessel function of the first kind of zero order

J_1 = Bessel function of the first kind of first order

Y_0 = Bessel function of the second kind of zero order

Y_1 = Bessel function of the second kind of first order

$\alpha_1 \alpha_1$ = roots of the Bessel function which satisfy

$$J_1\left(\alpha n\right)Y_0\left(\alpha n\right) - Y_1\left(\alpha n\right)J_0\left(\alpha n\right) = 0$$

$$T_r = \frac{c_{vr}}{d_e^2} = \text{time factor for radial flow and}$$

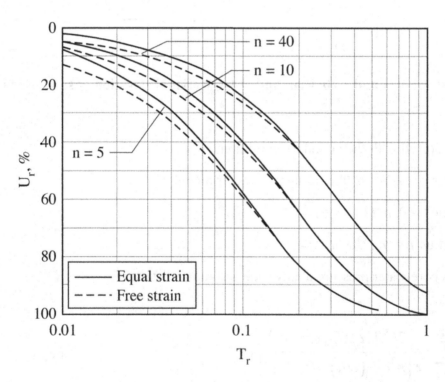

Figure 1.8 Consolidation as function of time factor (Richart, F. E.1959)

$$c_{vr} = m_v \gamma_w$$

k_h = coefficient of permeability in the horizontal direction.

The average degree of consolidation U_r is plotted as function of the time factor T_r in **Figure 1.8** where the ratio of the equivalent spacing and the drain well $n = r_e / r_w$.

1.8 Construction of sand drains

One of the most effective techniques of acceleration settlement caused by consolidation is the use of vertical sand drains with pre-loading. The sand drains can be installed by several techniques, such as:

(a) Jetted: the soil inside a driven pipe is removed by water jetted.

(b) Driven or vibratory closed-end mandrel: a closed steel casing with a detachable shoe is driven in the soil. The tube is filled with sand and then it is extracted leaving the sand in place.

(c) Hollow stem continuous flight auger: an auger is screwed down to the desired location, while the sand is injected as the auger is extracted. This method is recommended to minimize the effects of smearing of the sand drain, causing reduction of the permeability, less heave and less excess pore water pressure in the area of four diameters around the sand drain.

Example 1.3

The borehole profile of reclaimed land where it is proposed to build a new highway embankment 8 ft. (2.44 m) high is presented in Figure 1.9. The water table is at a depth of 2 ft. below the ground surface. The fill consists of a sandclay with $\gamma_d = 120 \, \text{pcf}(18.89 \, \text{kN} / \text{m}^3)$. Determine the maximum shear stress caused by the weight of the fill at a depth of 17 ft, and the amount of overstressing at this depth and the shear strength after consolidation is 90% complete. The shear strength varies with the depth as shown in Figure 1.10

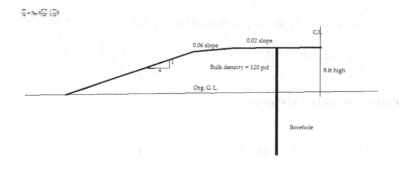

Figure 1.9 Embankment cross section

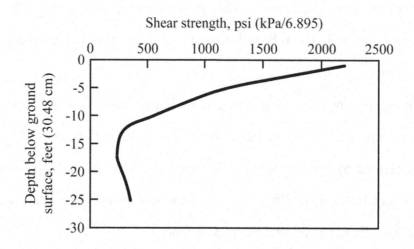

Figure 1.10 Shear strength as a function of depth

Solution

Since the ground is level before construction of the embankment fill, the shear stress is zero. The added weight of the embankment is p = 8(110) = 880 psf (2886 kN/m^2) maximum. After construction the maximum shear stress = p / π = 305 psf (1000 kN/m^2). Therefore the soil at depth = 17 ft (5.18 m) would be overstressed by (305—238)/238 = 28%. After 90% consolidation, shear strength would increase by 0.9(0.25) (1019) = 229 psf (751 kN/m^2). Since the original shear strength is 238 psf (781 kN/m^2), the increased shear strength would be 477 psf (1565 kN/m^2)or the factor-of-safety would be 1.56. ← **Ans.**

Example 1.4

A coastal highway is to be constructed by widening an abandoned railway as shown in Fig, 1.11. The subsoil consists of reclaimed land from the ocean which has been heavily desiccated by the sun, such that the shear strength of the soil decreases rapidly with depth and then increases again as a normally consolidated soil should. In order to accelerate the consolidation under the new added embankment, it is decided to use sand drains instead of surcharging the area over the new fill. Determine the spacing of 18 in. (45 cm) diameter wells so that 90% of the primary consolidation would occur in about 1 year, in time for completion of the subgrade preparation for the new highway. Calculate the time for 90% consolidation for $c_{vr} = c_v = 16 \, \text{ft}^2 / \text{yr}. \, (14,864 \, (\text{cm}^2 / \text{yr}) =$ the coefficient of consolidation radially and vertically.

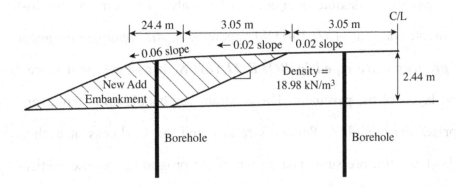

Figure 1.11 New added embankment with sand drains

Solution

Place a coarse sand blanket over the new fill before it is placed and compacted. Use sand drains 10 ft. apart as a trial.

$$\text{The time factor } T_r = \frac{16t}{10^2} = 0.16 \, t$$

For no smear

$$n = \frac{5}{0.75} = 6.67$$

17

From Fig. 1.8, for 90% consolidation T_r = 0.28. Therefore the time for consolidation is 1.75 yr. The time is too long. Try 8 ft. (2.44 m) spacing of the sand drains.

$$\text{The time factor } T_r = \frac{16t}{8^2} = 0.25\, t$$

Therefore the required time for 90% consolidation is 1.12 yr. This is close to the specified time. Adopt sand drains 8 ft. spacing, 18 in. in diameter for the design.

Without sand drains the time factor for 90 % consolidation = 0.848 and the corresponding time is 5.3 yrs. or five times slower than with sand drains.

1.9 Stress distribution in pavements

CHEVRON Multi-Layer Elastic Stress Analysis

Any type of pavement, flexible or rigid, can be analyzed by Chevron multi-layer stress analysis computer program *CHEVRON* (Appendix I). *All computer programs and only computer programs are typed in BOLD ITALICS.* It assumes that there is no bond between the layers of the pavement. The program can handle up to ten layers. The input data comprises the moduli E, Poisson's ratio v and the thickness of each layer H, the vehicular load and tire pressure. The output of the program gives the vertical σ_z, radial σ_r and tangential σ_r and shear stresses τ_{rz} and the corresponding strains. Figure 1.12a and Figure 1.12b show the cross section of concrete pavement and the distribution of the vertical normal and tangential stresses with depth below the surface, respectively. The corresponding diagrams for an asphalt concrete (AC) pavement are presented in Figure 1.13a and Figure 1.13b, respectively. Note the different magnitude of the stresses in the base and subbase layers of the two types of pavement.

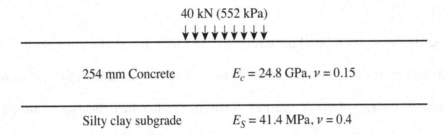

40 kN (552 kPa)

| 254 mm Concrete | $E_c = 24.8$ GPa, $v = 0.15$ |
| Silty clay subgrade | $E_S = 41.4$ MPa, $v = 0.4$ |

Figure 1.12a Concrete pavement cross-section

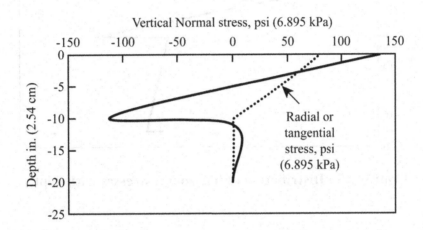

Figure 1.12b Distribution of radial and tangential stresses with depth

40 kN (552 kPa)

AC surface	$E_1 = 862$ MPa, $v_1 = 0.35$	$H_1 = 152$ mm
Gravel base	$E_2 = 276$ MPa, $v_2 = 0.35$	$H_2 = 305$ mm
Subgrade	$E_3 = 62$ MPa, $v_3 = 0.4$	

Fig 1.13a Flexible pavement cross section

Example 1.5

Determine the maximum bending stress and the maximum shear stress at a point 8 inches (20 cm) from the center of a 9000 lb. (40-kN) wheel load with tire pressure 80 psi (552 kPa) in the interior of the AC surface. Determine also the deflection of the pavement under the wheel load. The material properties of the paving layers are given below.

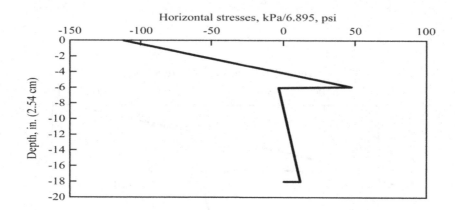

Figure 1.13b Distribution of horizontal stresses with depth

AC surface:	Dynamic modulus E_c^*	= 410 ksi (2827 MPa)
	Thickness	= 6 in (15.24 cm)
Subgrade: Gravel	Dynamic modulus	= 45 ksi (310 MPa)
	Thickness	= 12 in. (30.48 cm)

Solution

The maximum bending stress occurs at the underside of the AC layer under the center of the wheel load, and the maximum shear stress occurs just below the top of the AC surface layer close to the tire edge. The AC dynamic modulus in bending $\approx 0.33\ E_c^*$ is used to compute the maximum bending and shear stresses. From ***CHEVRON,*** the maximum bending stress = 12.80 psi (88.3 kPa), and the maximum shear stress = 9.40 psi (64.8MPa). The deflection of the pavement under the wheel load is 0.0107 in. (0.027 cm) The bending stress distribution the bottom of the AC is shown below ← Ans

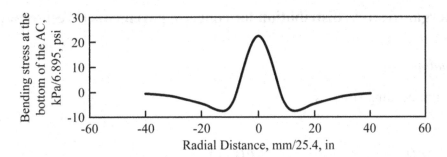

Figure 1.14 Bending stress at midpoint of the bottom of the AC surface

as a function of the distance of the wheel load from the point

1.10 Poisson's ratio

Poisson's ratio v is defined as the horizontal strain divided by the vertical strain. It is required for all elastic stress distribution. Fortunately, it is not a sensitive parameter, such that it may be assumed from book values for most practical purposes. Unless stated otherwise, the value may be taken as given in Table 1.2.

Table 1.2 Poisson's ratio

Material	Poisson's ratio
Asphalt Concrete	0.35
Portland cement concrete	0.15
EVAPAVE (new)	0.30
Sand	0.35
Silt	0.40
Silty clay	0.45
Clay	0.45-0.5

1.11 Westergaard stress distribution for concrete pavements (Westergaard, 1948)

Interior loading

The maximum bending stress is

$$f_b = \frac{3P(1+v)}{2\pi H^2}(ln\frac{2l}{a}+0.5\ -\gamma)\ +\frac{3P(1+v)}{64H^2}\frac{a}{l^2} \tag{1.10}$$

where P = load, H= thickness of slab, v = Poisson's ratio, a = the radius of the circular loaded area, $l = \sqrt[4]{(D/k)}$ = radius of relative stiffness of the slab, D = flexural rigidity = E, $H^3/(12(1-v^2))k$ = coefficient of subgrade reaction, and γ = 0.5772 = Euler's constant. The deflection at the interior of the slab under the load is

$$\delta_i = \frac{P}{8kl^2}\left[1+\frac{1}{2\pi}\left(\ln\frac{a}{2l}+\gamma-1.25\right)\right]\frac{a}{l^2} \tag{1.11}$$

Westergaard assumed an infinite slab in deriving Eq. (1.11). The results are sufficiently accurate if $L/l > 6$, where L = the smaller lateral dimension of the slab. The radius a of the loaded area should be small in comparison to l.

Edge Loading

Westergaard *defined edge loading when the circular wheel load is at the edge tangential to the edge, but at a considerable distance from any corner.* The maximum edge bending stress is

$$f_{be} = (3(1+v)\frac{P}{\pi(3+v)H^2}\{\ln\left(EH^3\right)/\left(100\,ka^4\right)+$$

$$1.84-4/3v+\frac{1-v}{2}+1.18(1+2v)\frac{a}{l}\} \tag{1.12}$$

The maximum deflection

$$\delta_e = \frac{P(2+1.2\mu)^{0.5}}{(EH^3k)^{0.5}}\left\{1-(0.76+0.4v)\frac{a}{l}\right\} \tag{1.13}$$

and the requirement for the development of the edge deflection is $L/l > 8$.

Corner loading

The distance of the point of maximum stress measured along the corner angle bisector is (Westergaard, 1926)

$$X_1 = 2(a_1 l)^{0.5} \tag{1.14}$$

where $a_1 = \sqrt{2}a$.

The corner deflection is:

$$\delta_c = \frac{P}{kl^2}(1.1 - 0.88(a_1/l)) \tag{1.15}$$

and the size requirement is $L/l > 5.0$.

For more stresses and deflection see also computer program **WESTRESS** (Appendix II)

Having designed a sound embankment and knowing how to calculate the stresses from vehicular loading, the engineer then turns his attention to the strength and deformation of the component layers including resistance to cracking (fatigue) and permanent deformation (rutting). Fatigue is handled by fracture mechanics and rutting by the stress-dilatancy theory.

PROBLEMS

1.1

It is proposed to construct a 4-lane coastal highway 19.5 m wide with an overall height of 1.5 m above the surrounding land comprising of alluvial soil deposits. The existing side slopes is 3:1.

Because of frequent flooding, it is proposed to construct a new highway over the old, with the same geometric design of the cross section. The shear strength profile with depth below the bottom of the embankment is as shown in Figure 1.2. Determine how high the new highway can be constructed so that the shear strength of the subsurface is not to be exceeded. Refer to standard embankment stress graphs in typical soil mechanics textbooks. Plot the extent of the overstressed zone.

1.2

For the rigid pavement shown in Figure 1.12a, calculate the design maximum bending stress with a 40 kN (552 kPa) wheel load (a) tangent to the longitudinal joint, (b) at the corner of the concrete pavement and (c) at the interior of the slab, according to Westergaard simplified theory.

Consider the pavement shown in Figure 1.13a, with the swelling potential of the subgrade shown in Figure 1.2. Examine the cost effectiveness on the swelling potential of the pavement of (a) doubling the modulus of the AC surface (b) doubling the thickness of the AC and (c) excavating 0.80 m of the expansive soil and replacing it with coarse sand. Comment on the total effects of each method.

1.3

For the flexible pavement shown in Figure 1.13a, calculate the design maximum bending stress for a 40 kN (552 kPa) wheel load at the base of the asphalt concrete, the vertical tangential strain and the maximum vertical displacement under the wheel load, the maximum shear stress in the pavement and where it occurs. The subgrade is usually assumed to infinitely deep. What difference would it make to the magnitude of the above parameters if the bedrock were at a depth of (a) 2.0 m and (b) 3.0 m?

1.4

Under what conditions would a flooded highway be (a) slightly damaged (b) badly damaged?

15

A flexible pavement consists of an AC layer with modulus $E_1^* = 600$, 1000, 1500 and 3100 MPa, a bituminous stabilized base with modulus of $E_2 = 500$, 700, 1000 and 1250 MPa, and a silty clay subgrade with modulus of 35, 50, 100 and 150 MPa. The Poisson's ratio may be taken from Appendix B. Plot (a) the variation of the maximum bending stress at the bottom of the AC base under a 40 kN wheel load with 552 kPa for (a) the variation of E_1^* with $E_2 = 700$ MPa (b) the variation of E_2 with $E_1^* = 1000$ MPa (c) the ratio of E_1^* / E_2 and (d) the variation with the modulus of the subgrade for $E_1^* = 900$ MPa, $E_2 = 500$ MPa. What practical significance can you infer from these plots?

1.6

The traffic comprises 80 kN axle loads. The pavement cross section is:

Concrete: 28-day compressive strength	34.47 MPa
Thickness	20 cm
Base course: modulus	800 MPa
Thickness	20 cm
Subgrade: modulus	62.5 MPa

Determine (a) the maximum edge bending stress (b) the maximum interior bending stress and (c) the interior deflection of the concrete slab.

1.7

Calculate the maximum edge bending stress for a concrete pavement with the following properties:

28-day compressive strength	37.92 MPa
Thickness of concrete	20 cm
Modulus of sandy gravel base	34.2 MPa
Thickness of base	20 cm
Modulus of sand subgrade	18.0 MPa

2

FRACTURE MECHANICS

"When you can define something and measure it, then you know something about it."

2.1 Griffith's Theory of Fracture

Griffith (1920) was the first to explain the observed wide discrepancy in the failure and calculated strengths of materials. He postulated that there was something in most materials that concentrates the relatively low applied stress in some region that causes the stress to become larger than the theoretical value of the fracture stress. His pioneering work in equating the strain energy released by cracking to that needed for the creation of new fracture surfaces at failure gave birth to the subject of *Fracture Mechanics*. Consider a metal plate of unit width containing a central crack of length $2c$ with a tensile stress σ applied at infinity, Griffith described the strain energy released when the crack propagates using Inglis' equation to determine the strain energy released as a crack grows as:

$$W_G = \frac{1 - \upsilon^2}{E} \pi \sigma^2 c^2 \qquad (2.1)$$

in which E = Young's modulus, and v = Poisson's ratio . The surface energy of a crack of length $2c$ and unit width is $4c\gamma$. Therefore the energy for its propagation is

$$W = 4c\gamma - \frac{1 - \upsilon^2}{E} \pi \sigma^2 c^2 \qquad (1.2)$$

Griffith theory of failure for brittle materials states that the crack will propagate spontaneously when $dW/dc = 0$, or

$$\frac{1-\upsilon^2}{E}\pi\sigma^2 c = 2\gamma \qquad (2.3)$$

The first quantity is defined as the strain energy release rate G named after Griffith. The critical value of the strain energy release rate is G_c. Griffith then formulated his theory of fracture of *brittle* materials for spontaneous growth of a crack, stating that *brittle failure occurs when the strain energy release rate becomes equal to its critical value* or

$$G = G_c \qquad (2.4)$$

in which G_c is a material property. Brittle materials are fracture sensitive. The more brittle the material the more fracture sensitive it is. *Fracture sensitivity is defined as the ratio of the stress at fracture of a notched specimen and that of the nominal stress based on the critical section.* It was not until 1960 that progress in the field of fracture was made, when Irwin (1960) introduced the concept of fracture mechanics.

2.1 Introduction to Fracture Mechanics

Irwin assumed that all brittle materials contain flaws or inherent cracks that vary in size from coalesced dislocations up to cracks that are visible by eye. He developed equations for the normal and shear stresses near the edge of a crack. In polar coordinates they are:

$$\sigma_y = K\frac{\cos\theta/2}{\sqrt{2\pi r}}(1+\sin\theta/2\sin3\theta/2)$$

$$\sigma_x = K\frac{\cos\theta/2}{\sqrt{2\pi r}}(1-\sin\theta/2\sin3\theta/2)$$

$$\tau_{xy} = K \frac{\sin\theta/2}{\sqrt{2\pi r}} \cos\theta/2 \cos 3\theta/2 \qquad (2.5)$$

in which K = *stress intensity factor,,*

$\tau_{yz} = \tau_{xz} = 0$, $\sigma_z = v(\sigma_y + \sigma_z)$ for plane strain $(\varepsilon_z = 0$ or constant$)$ and $\sigma_z = 0$ for plane stress.

There may be also a uniform stress parallel with the fracture surface.

The stress intensity factor K is defined as the parameter that governs the magnitude of the local stress field in accordance with Equations (2.5). It is the *most powerful parameter in all of engineering mechanics,* because it can describe the local stress field in terms of the applied stress, the crack size and the dimensions and shape of the body containing the crack. Its range of influence is the crack length measured from the crack tip (Figure 2.1). This is so because it is derived by nullifying the normal and shear stresses on the *crack surface in the uncracked pavement,* thereby creating the crack. There are hundreds of tabulated values of K by Tada et al. (1985) for most cases of practical interest. Cracks and discontinuities exert a dominant influence on pavement

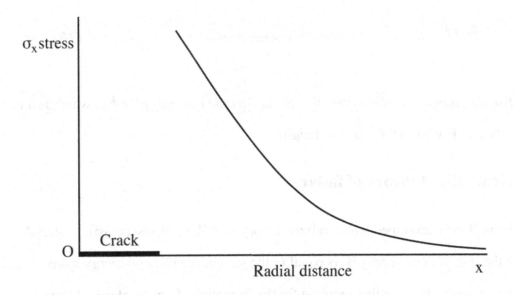

Figure 2.1. Stress in the vicinity of a crack tip

deformation. However, there are only three modes of fracture, as illustrated in Figure 2.2.

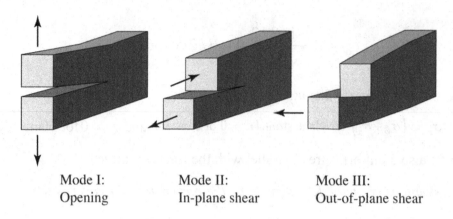

| Mode I: | Mode II: | Mode III: |
| Opening | In-plane shear | Out-of-plane shear |

Figure 2.2. The Modes of fracture in fracture mechanics

Irwin (1960) and Orowan modified Griffith's concept that the rate of strain energy release is equal to the surface energy to form new surfaces but also to include plastic flow and kinetic energy. The relationship between G and K is:

$$G = K^2 \frac{\left(1 - \upsilon^2\right)}{E} \qquad \text{for plane strain} \qquad (2.6)$$

$$G = K^2 \frac{1}{E} \qquad \text{for plane stress} \qquad (2.7)$$

The fracture toughness of a material is defined as the critical value K_c*, which is a material property that is vital in the study of fatigue.*

2.2 Generalized theory of failure

Williams (1965) developed a generalized theory of failure of viscoelastic materials, which states the rate of input energy $\dot{W}$ is equal to the sum of the rates of energy stored $\dot{W}_s$ and the energy dissipated $\dot{W}_d$ plus that required for the formation of new surfaces 4γ or

$$\dot{W} = \dot{W}_s + D + 4c\gamma \qquad (2.8)$$

Therefore
$$\dot{W} = \dot{W}_s + \dot{W}_d + 4\gamma \qquad (2.9)$$

where the dot refers to d/dc.

For a crack to grow spontaneously

$$\dot{W} = 0, so\ that\ \dot{W}_s \frac{dW}{dc} = G = 2\gamma \qquad (2.10)$$

2.3 Relationship between cracking and deflection

Consider a highway with a load P applied over a circular area on the pavement which contains a symmetrical crack of length $2c$ beneath the load. Let the deflection under the load be δ. Then the elastic energy is

$$W = \frac{1}{2}P\delta \qquad (2.11)$$

At the top of the loading cycle the rate of input energy is zero, so that the stored elastic energy can cause the crack to advance an incremental amount δc per unit area. Then

$$\frac{\partial W}{\partial c} = \frac{1}{2}P\frac{\partial \delta}{\partial c} \qquad (2.12)$$

$$K^2\frac{\left(1-\upsilon^2\right)}{E} = \frac{1}{2}P\frac{\partial \delta}{H\partial c} \qquad (2.13)$$

where H = the thickness of the slab. Therefore the increment in the deflection corresponding to the increment in the crack is

$$\frac{d\delta}{dc} = 2\frac{K^2}{PE}\left(1-v^2\right)H \qquad (2.14)$$

Thus the change in the deflection caused by the change in the crack length is

$$\delta = 4\int_0^{c_i}K^2\frac{\left(1-\upsilon^2\right)}{PE}Hdc + 4\int_0^{c_i}K^2\frac{\left(1-\upsilon^2\right)}{PE}Hdc \qquad (2.15)$$

where c_l and c_t = the half-crack lengths in the longitudinal and transverse crack, respectively. A computer program **DEFCRACK** (Appendix III) gives the relationship for the change in the deflection as a function of the increment in the crack length.

Example 3.1

Fatigue cracks caused by traffic are of 4 ft. (127 cm) long in the transverse and 8 ft. (101.6 cm) in the longitudinal directions, respectively. Calculate the deflection and the change in the deflection of the pavement shown in **Figure 1.3**, and the corresponding equivalent complex modulus of the cracked AC, assuming that it behaves like a homogeneous layer.

Solution

From **CHEVRON**, the displacement at the surface before the cracks developed is 0.014 in. (0.027 cm). From the computer program **DEFCRACK**, the deflection from the transverse crack and the crack 0.0107 in. (0.027 cm). Since the magnitude of the complex modulus is inversely proportional to the deflection, the equivalent modulus of the cracked AC surface is given by

$$E_{ec}^* = E^* \frac{\delta}{\delta_c}$$

where δ and δ_c = the deflection of the AC before and after cracking, respectively. From **DEFCRACK**

$$E_{ec}^* = 145000 \frac{0.0107}{0.014} = 110,000 \text{ psi} (764 \text{ MPa}) \leftarrow \textbf{Ans}$$

PROBLEMS

2.1 For an elastic material 1" thick loaded at infinity with a tensile stress σ, deduce Griffith's theory of fracture from William's generalized theory of failure.

3

STRESSES AT JOINTS AND CRACKS IN PAVEMENTS

The determination of the stress intensity factors in highway and airport pavements under realistic vehicular loading is the single most important factor in the application of fracture mechanics to the fatigue and fracture of such pavements. Folias et al. (1963) presented a theoretical solution for the SIF of a semi-infinite crack in a pavement subjected to general loading described by a Fourier series. This was followed by a theoretical solution for the SIF of a crack in a pavement subjected to a constant moment (Folias, 1970). However, in a pavement the cracks are too short to be considered semi-infinite and the stresses on the crack surfaces (in the uncracked slab) are far from constant, so that a more general solution is needed. Finite element methods are time consuming and expensive and not within the grasp of the practicing engineer. However, the weight function method is simpler, and offers some advantages over other analytical methods. This method is applied to determine the SIFs for cracks subjected to moving vehicular loads along the line of crack propagation. The analysis uses the subgrade modulus k as the parameter representing the elastic foundation.

3.1 Subgrade modulus for a plate on elastic foundation

Timoshenko and Woinowsky-Kreiger (1959) obtained an analytic solution for the bending stress and displacement for a *Plate Resting on a Semi-infinite Elastic Solid. A constant* k_0 *is defined as*

$$k_0 = \frac{E_0}{2\left(1-v_0^2\right)} \tag{3.1}$$

where E_0 and v_0 are Young's modulus and Poisson's ratio, respectively. The ratio of the maximum vertical pressure and the corresponding displacement was shown to be equal to k_0, a constant given by

$$\frac{p_{max}}{w_{max}} = k_0 \tag{3.2}$$

Evidently k_0 is the modulus of subgrade reaction for a two-layer elastic pavement. For a multilayered pavement the value of the subgrade modulus k has not been found analytically. There are numerous empirical expressions for the value of k such as the following for a two-layered system by Vesic and Saxena (1977)

$$k = \sqrt[3]{\frac{E_s}{E_c}} \frac{E_s}{H\left(1-v^2\right)} \tag{3.3}$$

where E_c and E_s = the moduli for the slab and subgrade, respectively. For a multilayer system, *the value of the subgrade modulus can be obtained directly from **CHEVRON** as the pressure divided by the displacement at the bottom of the surface layer directly beneath the wheel load.*

3.2 The stress intensity factors in a pavement slab with a semi-infinitely long crack

When normalized the crack length $\lambda' = \lambda c > 3.5$ the crack is considered to be *long*. Classical Kirchoff bending solutions for a normally loaded elastically supported flat plate containing a semi-infinite straight crack have been obtained by Ang et al. (1963). Consider an infinite plate on an elastic foundation with a through crack of length 2*c* as shown in Figure 3.1 (a) and (b).

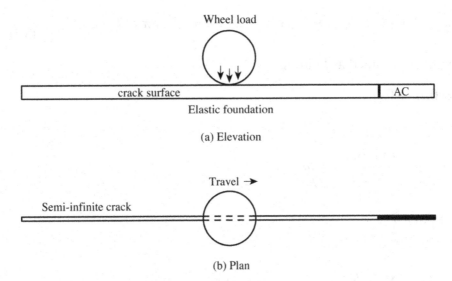

Figure 3.1 Wheel load on an elastic foundation

For a constant bending moment over the crack surfaces

$$M_y = -D m_0 e^{iax} \qquad (3.4)$$

where m_0 is a complex constant. Three limiting cases were presented for $c > 3.5/\lambda$. Ang et al. (1963) derived the value of the SIF as

Case 1. $\lambda' \neq 0, a \to 0 \left(a = \dfrac{2n\pi}{T} \right)$

$$K_{I0} = \left(\frac{3+v}{1-v} \right)^{0.5} \frac{EH}{2(1+v)\sqrt{2\lambda'}} m_0 \qquad (3.5)$$

Case 2. $\lambda' \to 0, a \neq 0$

$$K_{In} = -\frac{(1-i)EH}{(1-v)\sqrt{a}} \qquad (3.6)$$

Case 3. $\lambda' \to 0, a \to 0$. Ang *et al.* (1963) were unable to recover this limit from their solution.

Ramsamooj (1993) presented an approximation for this case. The residual moment is expanded into a Fourier series giving

$$M_y = \sum_{-\infty}^{\infty} -m_n e^{i2n\pi/T} = a_0 + \sum_1^n a_n \cos 2n\pi x / T + b_n \sin 2n\pi x / T \qquad (3.7)$$

The SIFs are then obtained as follows:

Case 1. $\lambda' \neq 0, a \rightarrow 0$

$$K_{I0} = \sqrt{\left(\frac{(3+v)(1-v)}{2\lambda'}\right)} a_0 \qquad (3.8)$$

Case 2. $\lambda' \rightarrow 0, a \neq 0$

$$K_{In} = \sum_1^{\infty} \frac{a_n + b_n}{\sqrt{\left(\frac{2n\pi}{T}\right)}} \qquad (3.9)$$

Case 3. $\lambda' \rightarrow 0, a \rightarrow 0$

When $a \rightarrow 0$ ($a \ll 1$, so that $n \ll T/2\pi$), the SIF in Eq. (3.6) must approach a limiting value in Eq. (3.7) for $a_0 = a_n + b_n$. The SIF is approximated by

$$\frac{1}{\Delta K_{In}} + \frac{1}{\Delta K_{I0}} + \frac{1}{\Delta K_{I0}} \qquad \textbf{(3.10)}$$

where $\Delta K_{I0} \sqrt{\left(\frac{(3+v)(1-v)}{2\lambda'}\right)}$ **and** $\Delta K_{In} + \sqrt{\left(\frac{T}{2n\pi}\right)} \qquad (3.11)$

Therefore

$$K_I = \sum_1^n \left(\Delta K_I = \sum_1^n \left(\Delta K_I (a_n + b_n) \quad K_I (a_n + b_n)\right)\right) \qquad (3.12)$$

For $1/\lambda < c < 3.5/\lambda$, the values are obtained by interpolation (Ramsamooj, 1993).

3.3 Stress intensity factor for a plate on an elastic foundation containing a crack

Folias (1970) considered an infinite plate on an elastic foundation with a through crack of length $2c$, symmetrically loaded by a constant bending moment over the crack surfaces. He obtained the SIF For $\lambda < c$, Folias obtained the SIF as:

$$K_I = f_b \sqrt{\lambda c}$$

$$(3.13)$$

where $\sigma = f_b$ = bending stress. For $1 < \lambda c < 3.5$, the value of K_I is given by (Figure 4.2) as:

$$K_I \frac{f_b \sqrt{c\pi}}{1 + a\lambda^2 c^2}$$

$$(3.14)$$

where $a \approx 0.125$. K_I is used as a reference K_{Iref}. to obtain the SIF for any other symmetrically loaded system as discussed below.

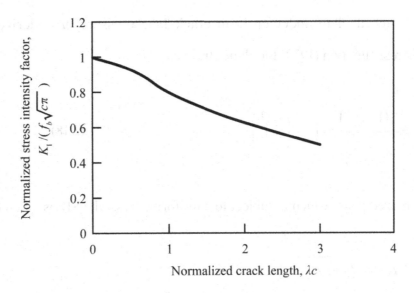

Figure 3.2 Stress intensity factor vs. crack length plate on elastic foundation

3.4 Stress intensity factors by the weight function method

Beukner (1970) proposed the weight function method to find the stress intensity factor

$$K_I = \int \sigma(x) h(c, x)) dx$$

$$(3.15)$$

where $\sigma(x)$ is the stress distribution across the plane of the crack in the uncracked body, loaded by the force system for which the K-value is being determined, x = the coordinate distance measured from the center of the crack along the crack, and h = *weight function defined by*

$$h(c,x) = \frac{E}{K_{Iref}\left(1-v^2\right)}\frac{\partial u}{\partial c} \tag{3.16}$$

where E = Young's modulus, V = Poisson's ratio, and u = the crack opening displacement function. Eq. (4.15) must reduce to an identity if the stress is exactly equal to that of the reference bending stress, or

$$K_{Iref}^2 = \frac{E}{1-v^2}\int_0^c \sigma(x)\frac{\partial u}{\partial c} \tag{3.17}$$

For a center cracked plate under uniform crack face pressure, u is derived from the Westergaard stress function (1939) for plane strain as

$$u(c,x) = \frac{2\left(1-v^2\right)}{E}\sigma c\left(1-\left(\frac{x}{c}\right)^2\right)^{1/2} \tag{3.18}$$

For a center cracked plate, which is subject to a uniform stress, the SIF is given by

$$K_I = f_b\sqrt{c\pi} \tag{3.19}$$

By analogy with Eq. (4.14), it is deduced that the crack opening displacement for the plate on an elastic foundation is

$$u(c,x) = \frac{2\left(1-v^2\right)}{E}\frac{\sigma c}{1+a\lambda^2 c^2}\sqrt{\left(1-\left(\frac{x}{c}\right)^2\right)} \tag{3.20}$$

Performing the computation in Eq. (4.14), the dimensionless weight function is

$$\Omega = h\sqrt{c} = \frac{2\left(1 - a\lambda^2 c^2 + 2a\lambda^2 x^2\right)}{\left(1 + a\lambda^2 c^2\right)\sqrt{\pi(1-(x/c))}} \tag{3.21}$$

Therefore, the Mode I SIF a specified mode of stress on the crack faces may be obtained from Eq. (3.15). The integration is done across the plane of the crack in the uncracked body.

Various types of loading may be superposed. The principle of superposition for linear elasticity implies that for the purposes of calculating SIFs, loading the crack faces with $\sigma(x)$ is equivalent to loading the cracked body with loads which give rise to $\sigma(x)$ in the absence of a crack (Petroski and Achenbach, 1978). Gorner et al. (1985) pointed out that the reference system used in the weight function method cannot be chosen arbitrarily, but should be uniform. Niu and Glinka clarified this point. They found that the reference system need not be restricted to a uniform stress, providing that

(1). the crack surfaces are fully loaded over the entire length, and

(2). the stress is a continuous function that varies monotonically without increase along the crack length.

It was further noted, that the stress gradient does not affect the accuracy of the results. The loading on the crack surfaces for which the SIF is required must the symmetrical with respect to the crack line for Mode I, but it does not have to be symmetrical with respect to the mid-normal to the crack (Petroski and Achenbach, 1978, Niu, 1990).

3.5 Deflection at cracks and joints in pavements

The first step is to determine the stresses at joints and cracks in the uncracked pavement using **CHEVRON**. Substitution of the respective functions into Eq. (3.15) and numerical integration by the computer program **EFM** (App. IV) gives the values of, if the loading is symmetrical with respect to the line of the crack. If the loading is not symmetrical with respect to the line of the crack, the stresses on the crack surfaces in the uncracked pavement consists of both normal and shear stresses. The cracking modes of interest in highway pavement engineering are primarily Mode I, the opening mode and the out-of-plane tearing Mode III (Fig. 2.2b). By analogy (Sih and Liebowitz, 1968), for plane strain or plane stress

$$K_{\mathrm{III}} = \int_0^c \tau(x) h(c,\ x) \mathrm{d}x \qquad (3.22)$$

where $\tau(x)$ = the shear stress.

For *combined Modes I and III* of crack propagation from the maximum principal stress theory (De Chang et al., 1978), the equivalent SIF is

$$K_{Ieq} = \frac{1}{2} K_I (1 - 2\upsilon) + \frac{1}{2} (K_I^2 (1 - 2\upsilon^2) + 4 K_{III}^2) \qquad (3.23)$$

Knowing the equivalent SIF for any crack of interest, the change in the deflection of highway or airport pavement, flexible or rigid, under the wheel loads at a joint or crack is obtained from **Eq. (2.14)** as

$$w_P = \int_0^c \frac{K_{Ieq}^2 (1 - \upsilon) H}{PE^*} \ \mathrm{d}c \qquad (3.24)$$

For concrete pavements the stresses may be obtained from the theory of an infinite plate on an elastic foundation or from **CHEVRON**. The joints are treated as cracks in an infinite slab. The effect of each joint on the deflection of the pavement is determined separately and then superposed to obtain the final deflection.

3.6 Deflection at point Q due to load at P

The deflection at point Q caused by a wheel load at point P in the position shown in Fig. 3.3b can be obtained from the reciprocal theorem in elasticity, which states that the deflection w_{PQ} at P due to the wheel load at Q, is equal to the deflection w_{PQ} at Q due to the load at P

Since a crack can be formed merely by superposition of equal and opposite stresses in the uncracked pavement, the principle of superposition is applicable. In fact it is applicable to any elastic system, even with cracks and holes (Timoshenko and Woinowsky-Kreiger, 1959). When equal loads are applied at both P and P', and at Q and Q'

$$w_P + w_Q + 2w_{QP} = \frac{2(1-\upsilon^2)}{PE} \int_0^1 (K_{IP} + K_{IQ})^2 H \, dc \qquad (3.25)$$

in which K_{IP} and are the SIFs for the crack due to the loads at P and P' and at Q and Q, respectively.

Accordingly the deflection profile of the concrete slab along the midnormal to the crack can be found. A wheel load positioned at a point Q along the midnormal to the crack as shown in Figure 3.3b, causes a bending stress as well as a shear stress along the crack surface

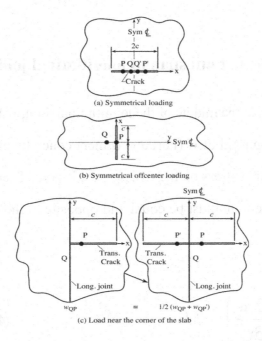

(a) Symmetrical loading

(b) Symmetrical offcenter loading

(c) Load near the corner of the slab

Figure 3.3 Loading for cracks or joints in an infinite slab on an elastic foundation

This generates deformation in both Modes I and III. The values of K_I and K_{III} are determined by Eqs. (4.15) and (4.22), K_{IeqQ}, respectively. The equivalent Mode I SIFs obtained for the load

P and Q are denoted K_{IeqP} and K_{IeqQ}, respectively. The deflection at point P caused by the load at P_w is

$$w_{\text{P}} = \frac{4(1 - \upsilon^2)}{P_w E} \int_0^1 K_{\text{P}}^2 \, H \, dc \qquad (4.27)$$

where l = the half crack length. The deflection at point Q caused by the load at P is

$$w_{\text{QP}} = \frac{4(1 - \upsilon^2)}{P_w E} \int_0^1 K_{\text{IeqP}} \, K_{\text{IeqQ}} H \, dc \qquad (4.28)$$

When the point Q is located as shown in Figure 4.3c, the deflection is given by half of the sum of the deflections caused by two equal loads positioned on either side of the point. The method of combining **Elastic Theory** and **Fracture Mechanics** to obtain the deflection at any point in terms of the SIFs is called EFM. The computer program is also called ***EFM*** (Appendix IV).

4.7 Edge bending stress at midspan of longitudinal joint

The tensile bending stress at the middle of the longitudinal joint, with the load tangent to the joint, is used for designing rigid PCC (Portland cement concrete) pavements. The boundary condition is that the bending stress on the free surface is zero. Therefore the bending stress due to the presence of the joint must be equal and opposite to those caused by the applied load, so that

$$D\left(\frac{\partial^2 w}{\partial y^2}\right) + \upsilon\left(\frac{\partial^2 w}{\partial x^2}\right) = M_y \qquad (4.29)$$

where M_y = the bending moment caused by the primary loading. The second derivative $\partial^2 w / \partial x^2$ is obtained from the deflection profile along the edge using finite difference, or

$$\frac{\partial^2 w}{\partial x^2} = 2\frac{w_2 - w_1}{\Delta x^2} \qquad (4.30)$$

where W_1 and are the deflection at the mid-slab position and at a Δx distance from it. The bending moment in the x-direction at the mid-slab position is

$$-D\left(\frac{\partial^2 w}{\partial x^2} + \upsilon\frac{\partial^2 w}{\partial y^2}\right) = -M_x \qquad (4.31)$$

This is the bending moment required for design.

As pointed out by Huang (1993), in finite element computations, it is necessary to multiply the *k*-value used for interior stresses by a factor of 1.75 in order to obtain satisfactory agreement with Westergaard theory for the edge stresses. However, there is no rationale for this. No adjustment is necessary when using *EFM* (IV)

4.8 The computer program *EFM*

Input Data

1. Young's modulus and Poisson's ratio for each layer

2. The thickness of each layer

3. The length and width of the slab

4. The magnitude of the axle load and location of the wheels

5. The points where the shear and bending stresses and deflection are required.

Computer Output

1. The stresses and deflection in the unjointed pavement at the specified points are obtained from *CHEVRON*.

2. The stresses and deflection at the specified points caused by the presence of the joints, and

3. The combined stresses and deflection by superposition of the above.

Example 3.1

Find the deflection and edge bending stresses for typical PCC pavements shown below. In all cases the length of the slabs is 5.90 m.

Young's modulus, E_c GPa	20.68	34.47	27.58
Subgrade reaction k, MPa	13.79	68.95	31.03
Slab thickness H, cm	20.3	356	254

Solution

The subgrade reaction for a two-layer system by (Vesic and Saxena, 1974)

$$k = \sqrt[3]{\frac{E_s}{E_c}} \frac{E_s}{H\left(1 - \upsilon^2\right)}$$

The value of k for deflection was taken 0.42 k for stresses given above. The results for the tensile stress at the bottom of the concrete and deflection were obtained for full factorial combination of the above parameters:

The results for the bending stresses at the edge of the pavement with the wheel located tangentially to the longitudinal edge midway between the transverse joints, obtained from EFM theory are presented in Figure 4.3, together with those determined by the Westergaard theory and by finite element computations by Ioannides et al. (1985) and by Darter (1977). There is good agreement in all cases.

Example 3.2

Find the bending stress and deflection at the mid-slab position in the longitudinal edge.

Solution

Concrete pavements of varying thicknesses were analyzed to obtain the bending stress and deflection at the mid-slab position in the longitudinal edge. The results for the bending stresses and deflection obtained by EFM are presented in Figures 3.3, 3.4 and 3.5, together with the experimental data from the AASHO Road Test (1972) for the same pavement with the axle load in the same location, showing good agreement.

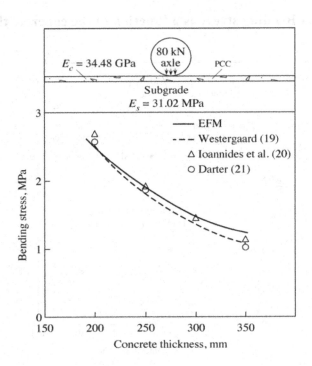

Figure 3.3 Edge bending stress as a function of the concrete thickness

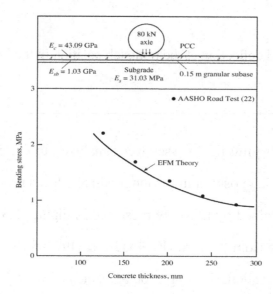

Figure 3.4 Bending stress as a function of the concrete thickness

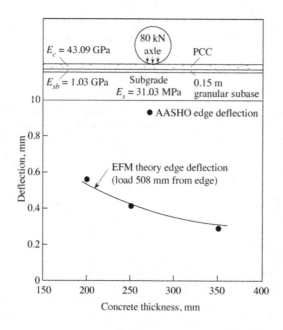

Figure 3.5. Edge deflection as a function of the concrete thickness

Measurements of the corner deflection in the AASHO Road Test were made in the location shown in Fig.3. 6. The results computed by EFM theory, together with the experimental data are compared in Fig.3. 7. The measured deflections were very sensitive to the time of the day, because of warping caused by temperature gradients. The curling of the concrete was assumed to be zero, when the temperature at 6.3 mm below the top of surface was equal to the ambient temperature.

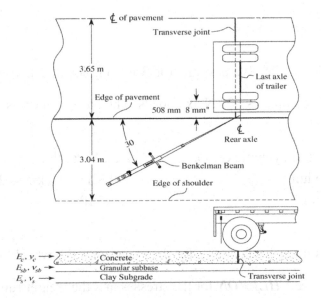

Figure 3.6 Full scale measurement of the corner deflection at the AASHO road Test

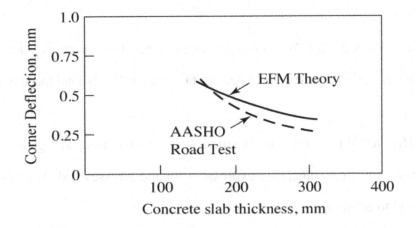

Figure 3.7. Comparison of the corner deflection by

EFM theory and in the AASHO Road test

The agreement between EFM predictions and the experimental data is reasonable considering the large experimental variation with a standard deviation of 37%.

PROBLEMS

3.1

A longitudinal crack in a flexible pavement is 3 m long located 1 m from the edge of the pavement. The traffic comprises 80 kN axle loads. The pavement cross section is:

Surface: modulus in dynamic comp.	1722 MPa; thickness 20 cm
Base course: modulus	800 MPa; thickness 20 cm
Subgrade: modulus	62.5 MPa

Determine the stress intensity factor K_I for the crack with the wheel load directly over the middle of the crack. Use **CHEVRON** for the stresses and the weight function method for K_I

3.2

The pavement of Prob. 3.1 the AC contains a longitudinal through-crack 6 feet long, 3 feet from the shoulders. Calculate the stress intensity factor for a 40 kN load located at the middle of the crack.

Hint: Use **CHEVRON** to obtain the bending stress distribution along the crack in the uncracked pavement, fit an algebraic expression for the stresses and then use the weight function method to compute the SIF

4

FRACTURES OF HIGHWAY AND AIRPORT PAVEMENTS

Stress intensity factors for pavements containing a crack

The general method of solution for the SIFs is to determine the stresses in the uncracked pavement and then to nullify the stresses over the crack surfaces where the crack is desired. Suppose that a pavement slab has a symmetrically loaded crack of length $2c$ for which the SIF is needed. In the classical theory of bending of thin plates, it is possible to satisfy stress-free conditions at an edge in an approximate way by requiring that the bending moment and shear along the crack vanish. For the uncracked slab, the moment and shear along the crack surface can be made to vanish by requiring that the homogeneous solution, providing that it satisfied certain physical condition far away from the crack, must be equal and opposite to the particular solution.

4.2 Formulation of the problem of crack in a pavement loaded symmetrically

Introducing non-dimensional coordinates $x = X/c$ and $y = Y/c$, the homogenous part of Eq. (5.1) becomes

$$\nabla^4 w(x,y) + \lambda c^2 w(x,y) = 0 \qquad (4.1)$$

where $\lambda = \sqrt[4]{k/D}\,c$. The boundary conditions to ensure that the crack surface is stress free at $y = 0$ and $x < c$ is:

$$M_y(x,0) = -\frac{D}{c^2}\left(\frac{\partial^2 w}{\partial y^2} + v\frac{\partial^2 w}{\partial x^2}\right) = -M_y^p(x,0)\, x \le 1 \tag{4.2}$$

$$Q_y(x,0) + \frac{1}{c}\frac{\partial M_{xy}}{\partial x}(x,0) = -\frac{D}{c^2}\left(\frac{\partial^3 w}{\partial y^3} + \left(2-v\frac{\partial^3 w}{\partial x^3 \partial y}\right)\right) = 0\, x \le 1 \tag{4.3}$$

Because of symmetry (Figure 4.1)

(a) w must be even in x and y

(b) M_{xy} and Q_y must vanish as $y \longrightarrow 0$, not only on the crack, but for all x; and

(c) the slope must vanish as $y \longrightarrow 0$, for all x such that $x \le 1$.

The condition (b) and (c) can be shown to be equivalent to the requirement that w and its partial derivatives be continuous across $y = 0$, for all x such that $x \le 1$.

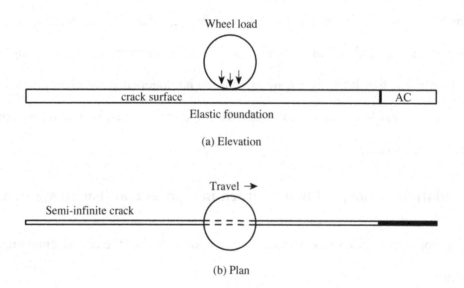

Wheel load

crack surface | AC

Elastic foundation

(a) Elevation

Travel →

Semi-infinite crack

(b) Plan

Figure 4.1 Crack in a pavement symmetrically loaded

The deflections may be represented by the following integral which is symmetrical with x and y:

$$w(x, y) = \int_0^\infty \{P_1(s)e^{-\lambda\sqrt{s^2-\alpha^2}} + P_2(s)e^{-\lambda\sqrt{s^2-\beta^2}}\} \cos \lambda xs \, ds \qquad (4.4)$$

where $P_1(s)$ and $P_2(s)$ are functions to be determined from the boundary conditions. Folias (1970) obtained them by integration. They are:

$$P_1(s) = -\frac{v_0 s^2 + \beta^2}{\sqrt{(s^2 - \alpha^2)}} \left(\frac{1}{s} A_1 J_1(\lambda s) + \frac{3\lambda}{s^2} A_2 J_2(\lambda s) + \ldots\right) \qquad (4.5)$$

$$P_2(s) = \frac{v_0 s^2 + \alpha^2}{\sqrt{(s^2 - \beta^2)}} \left(\frac{1}{s} A_1 J_1(\lambda s) + \frac{3\lambda}{s^2} A_2 J_2(\lambda s) + \ldots\right) \qquad (4.6)$$

The exact solution of Equations of this type is very difficult. Erdogan and Kebler (1969) used numerical methods to solve a similar equation. Folias (1970) obtained an approximate analytical solution of Eq. (4.4) for $\leq x \geq 1$ by integration. From Sih and Setzer (1964), the Kirchoff SIF for plate bending is defined as

$$K_{IB} = \frac{\sigma_{xx} + \sigma_{yy}}{2\cos\theta/2} \frac{3+v}{1+v} \sqrt{2\pi r} \qquad (4.7)$$

$$\sigma_{xx} + \sigma_{yy} = -\frac{Ez}{(1-v^2)} \left(\frac{\partial^2}{\partial x^2} + \frac{\partial^2}{\partial y^2}\right) w \qquad (4.8)$$

where w is given by Eq. (4.4).

Consider only the singular part of the solution arising from the expansion of the integrals for large s (Gradshteyn and Ryzhik, 1965),

$$(4.9))$$

$$\int_0^\infty e^{-\lambda s y} J_1(\lambda s) \cos \lambda xs \, ds = \frac{1}{\lambda} \sqrt{\left(\frac{c}{2r}\right)} \cos\theta/2$$

where r/c is the non-dimensional distance from the crack tip and $\theta = \tan^{-1} y/x$. Therefore

$$K_{IB} = -\frac{Eh}{2c^2} \alpha^2 \lambda \frac{3+v}{1+v} A_1 \sqrt{c\pi} \qquad (4.10)$$

51

Folias et al. (1972) derived the following analytical expression for a crack of length $2c$ loaded symmetrically with a bending stress f_b.

$$K_I = \frac{f_b \sqrt{c\pi}}{1 + a\lambda^2 c^2} \tag{4.11}$$

in which $a = 0.125$.

The value of the non-dimensional constant can be obtained from Eqs. (4.10) and (4.11). Eq. (4.11) gives the value of the SIF for a crack for the radius of relative stiffness $3.5 > \lambda \geq 1$. Under highway or airport loading long longitudinal cracks develop leading to fatigue distress and rapid deterioration of the pavements. The solution for the SIFS for long cracks under a moving load as shown in Figure 5.1 is useful in determining the behavior of a crack after it has reached its critical length.

4.3 Stress intensity factors for pavements with a long crack

Classical Kirchoff bending solutions for a normally loaded elastically supported flat plate containing a semi-infinite straight crack have been obtained by Ang et al. (1963). When c > 3.5/λ, the crack may be considered to be semi-infinite. From these solutions Sih and Setzer (1968) obtained the SIFs as follows. Let M_y be the residual bending moment along the crack surface that must be nullified to obtain the SIFs in the cracked plate such that

$$M_y = -Dm_0 e^{iax} \tag{4.12}$$

where m_0 is a complex constant. Three limiting cases were presented in Section 3.

In the derivations of the above results, the classical Kirchoff theory of bending was used. It is an approximate theory that accounts for the vanishing of the moment and shear and not the individual stresses. The use of the more accurate Reissner's theory gives the relation between the Kirchoff SIF K_{IB}, and the Reissner SIF K_I given (Hartranft and Sih, 1967) as

$$K_I = \phi K_{IB} \tag{4.13}$$

where is dependent on the parameter H/c and v, where H = the thickness, c = the half crack length and v is Poisson's ratio. For $v = 0.3$, ϕ is given by

$$\phi = 0.9 - 0.03\left(2 - 0.316\,H\,/\,c\right)^3 \tag{4.14}$$

4.4 Stress intensity factors for pavements with a semi-elliptical crack

The semi-elliptical crack has long been recognized as a close approximation to "a natural flaw" (Marrs and Smith, 1972). For a semi-elliptical or quarter-elliptical crack, with minor and major semi-axes a and c, respectively, the SIF at the leading horizontal edge of the crack tip at the bottom of the top layer of a pavement, is given by Grandt and Sinclair (1972)

$$K_I = \frac{M_B \sigma_b}{Q\sqrt{\pi a}} \tag{4.15}$$

where $M_B = 1.02 - 0.2\ a/h$ and h *is* the thickness of the pavement layer. An approximate value of Q is given by Broek (1967) as

$$Q = \left(1.18 + 0.39\left(\frac{a}{c}\right)^2 - 0.212\left(\frac{f_b}{f_r}\right)^2\right) \tag{4.16}$$

where σ_b = average bending stress, f_r = ultimate bending strength and a/c = shape factor. Since most natural flaws have a shape factor $a/c = 0.5$-0.3, it is assumed that the shape factor in the pavement is 0.4.

4.5 Application to fracture of pavements

In order to determine this criticality of a specific crack, it is first necessary to determine the SIFs. The crack in the interior of the pavement is assumed to grow from a semi-elliptical crack into part-through crack and eventually into a long crack $(\lambda' = \lambda c > 3.5)$. The dominant mode of crack growth is Mode I, because the shear stress is usually much smaller than the bending stress. The fracture toughness is an essential material property. The *ASTM criteria for fracture toughness determination experimentally are that the crack length, the width of the beam and the remaining length of the ligament above the crack should all be greater than twenty five times the size of the inelastic or plastic zone.*

The experimental values of K_{Ic} for asphalt concrete reported in the published research literature do not satisfy the ASTM criteria. The size of the beams tested were too small (typically 5" x 2.5"x 20" long) mainly because it is impracticable to fabricate beams large enough, the main problem being the size of the oven for heating up of the materials to 350 deg. F and then compacting it. The fracture toughness requirements in concrete are easier to meet, because no heat is involved. **Bazant (1993)** proposed an empirical formula for the prediction of the fracture toughness of concrete (PCC), on the basis of the tensile strength f_t, and the maximum aggregate size d_a, as:

$$K_{Ic} = 0.146 \sqrt{\left(\left(f_t + 127 \right) d_a \right)} f_t (\text{psi} \sqrt{\text{in}}) \qquad (4.17)$$

Because of the cost and difficulty in determining the fracture toughness of AC, it is necessary to use the same empirical expression as for PCC. For an AC f_t = 254 psi, K_{Ic} = 626 psi $\sqrt{\text{in}}$ $\left(682 \text{ kPa} \sqrt{\text{m}} \right)$. If the size of the inelastic zone is taken as that of the Irwin plastic zone (1.63 cm (0.64 in.)) then the minimum width of the AC beam for laboratory testing is 25 times the size of the plastic zone or 40 cm. The depth of the beam would be 80 cm and the length 320 cm.

The SIFs are now used to study the fracture of three types of pavements in the following.

Flexible Pavement

AC properties (Majidzadeh et al,1976)

Modulus of elasticity	= 1241 MPa
Poisson's ratio	= 0.35
Fracture toughness	= 682 kPa $\sqrt{m}$
Thickness	= 15.24 cm
Subgrade	
Modulus of elasticity	= 62 MPa
Poisson's ratio (assumed)	= 0.4

Rigid PCC pavement

Material properties from Darter (1977)

Modulus of elasticity	= 34,475 MPa
Poisson's ratio (assumed)	= 0.15
Fracture toughness (Eq. 55)	= 923 kPa $\sqrt{m}$
Thickness	= 25.24 cm
Tensile strength	= 1.32 MPa
Subgrade	
Mod. of subgrade reaction	= 54.1MPa
Poisson's ratio (assumed)	= 0.4

Bending stresses in the pavements

The bending stress at the bottom of the AC pavement surface layer was computed from **CHEVRON**. The corresponding values for the rigid pavement were obtained from Darter (1977) and those for the rigid airport pavement from Packard (1973).

Flexible highway pavement:

Bending stress at bottom directly beneath wheel load   =749 kPa

Rigid highway pavement:
    interior loading: 893 kPa
    thermal curl (0.43 deg C/cm)   1354 kPa

Rigid airport pavement:
    interior loading: 2391kPa
    thermal curl (0.43 deg. C/cm)   2027 kPa

4.6 Crack location in the pavements leading to fracture

Fracture occurs when the SIF at the tip of a growing crack under fatigue loading exceeds the fracture toughness of the material. For a typical flexible pavement the critical crack is in the longitudinal wheel path, but for a typical rigid pavements the critical crack is the transverse crack, due to greater overlap of the σ_y -stresses from the two wheel loads compared to that of the σ_x —stresses (Vesic, 1972).

Figure 4.2a shows that in a typical flexible pavement carrying a 24-K (106.7 kN) axle load, the maximum SIF is only $320 \, \text{kN} \, \text{m}^{1.5}$, so that the fracture toughness is not exceeded. Figure 4.2b shows the SIFs for a transverse and a longitudinal crack for a *PJCP highway pavement* for an 18 kip (80 kN) single axle load. Because of symmetry, the transverse crack length of interest is only 91.5 cm or one-quarter of the lane width. The maximum SIF is 2310 kPa $\sqrt{\text{m}}$, which should be increased by 20% for impact (Yoder, 1959).

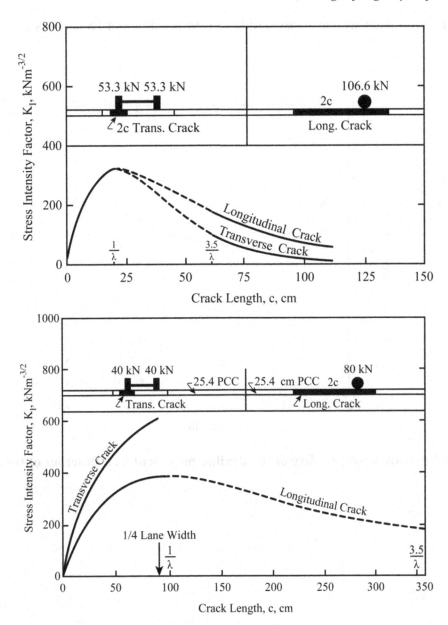

Figure 4.2a and Fig 4.2b. Stress intensity factors for AC and PCC highway pavements

This value is much greater than the fracture toughness of the concrete of 1225 kPa√m
. Therefore fracture will occur in both the transverse and longitudinal directions. If the
thermal curl stress is added to the load stress, the maximum SIF becomes 3603 kPa√m, and
the cracks will propagate from 30 to 420 cm in the longitudinal direction before stable crack
growth resumes. However, the thermal stress can be nullified by the moisture gradient stress
caused by the drying out of the top of the concrete slab.

Flexible pavements are most susceptible to fracture at about 15 degrees C as shown in Figure 4.5.

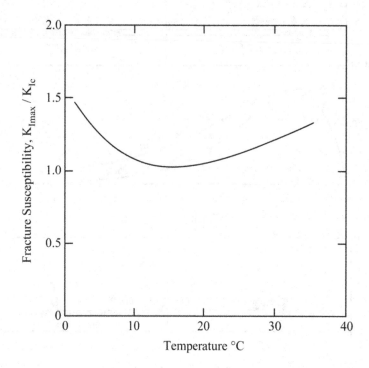

Figure 4.5 Fracture susceptibility of the flexible pavement as a function of temperature

5

VISCOELASTICITY

"In our quest for objectivity, simplicity, and precision, we customarily fit behavior to rigid mathematical models which make no provision for systems complexity and the imprecision in our cognition, perception, evaluation and decision-making processes"
Lee Young, Professor of Geography, Chinese University of Hong Kong, 1988

Viscoelasticity is the property of materials that exhibit both time dependent viscous and instantaneous elastic characteristics when undergoing deformation. The viscoelastic behavior is a result of molecular rearrangement. Some phenomena in viscoelastic materials are:

(a) when the stress is held constant, the strain increases with time (creep);

(b) when the strain is held constant, the stress decreases with time (relaxation);

(c) the stiffness and the time for failure depend on the rate of loading;

(d) during cyclic loading hysteresis occurs, causing loss of mechanical energy.

5.1 Mechanical models for viscoelastic materials

There are many mechanical models simulating the behavior of these materials. The Maxwell model consists of a spring and dashpot in series. A Kelvin model consists of a spring and a dashpot in parallel. The Burger's model combines the Maxwell and Kelvin models. It provides an excellent representation of the behavior of some materials in both loading and unloading. Mathematically it is described as follows:

$$\varepsilon = \frac{\sigma}{E_0} + \frac{1}{E_1}\left(1-\exp(-\frac{t}{T_1})\right)\sigma^n + \frac{t}{E_0 T_0}\sigma^n + C_p\sigma^n \tag{5.1}$$

If both n and p are unity, the first term is elastic, the second is anelastic or transient, the third is viscous, and the last is plastic. In Eq. (5.1), = the deviator stress; t = calendar time, E_0 = elastic modulus; T_0 = the relaxation time of the Maxwell model, which is the time for the relaxation to cause a reduction of the deformation to 36.8% of its initial value; T_1 = the retardation time of the Kelvin model, or the time for the retardation to reach 63.3% of the total time dependent strain; an C_p = constant. The creep compliance $J(t)$ defined as the strain per unit stress consists of the first three terms. For a linear viscoelastic material

$$J(t) = \frac{\varepsilon}{\sigma} = \frac{1}{E_0} + \frac{1}{E_1}\left(1-\exp(-\frac{t}{T_1})\right) + \frac{t}{E_0 T_0} \tag{5.2}$$

5.2 Cyclic Loading Response

Under cyclic loading, the linear viscoelastic deformation is given by the superposition integral as

$$\varepsilon(t) = \sigma(t)J(0) + \int_0^t \sigma(t)\frac{dJ(t-\tau)}{d(t-\tau)} \tag{5.3}$$

For a haversine load pulse on the linear viscoelastic material modeled above, the contribution of the transient term to the permanent deformation per cycle is

$$\varepsilon_y = \int_0^{\Delta t} \frac{\sigma}{2}\left\{1-\cos\left(2n\frac{t}{p}\right)\right\}\left\{\frac{\exp(-(t-\tau))/T_1}{E_1 T_1}\right\}d\tau \tag{5.4}$$

Therefore

$$\varepsilon_y = \frac{\sigma T_1}{2\left(p^2 + 4\pi^2 T_1^2\right)} \left\{ \frac{-p^2 - 4\pi^2 T_1^2 + p^2 \cos\left(\omega t_2 - 2p\pi \sin(\omega t_2)\right)}{\exp(t_1 / T_1)} \right\}$$

$$-\frac{-p^2 - 4\pi^2 T_1^2 + p^2 \cos\left(\omega t_1 - 2p\pi \sin(\omega t_1)\right)}{\exp(t_1 / T_1)} \tag{5.5}$$

where $\omega = \dfrac{2\pi}{p}$; and $, t_2$ = time at the start and the end of the cycle, respectively. The contribution of the viscous term to the time dependent deformation per cycle is

$$\varepsilon_y = \int_0^{\Delta t} \frac{\sigma}{2} \left\{ 1 - \cos\left(\frac{2\pi\tau}{p}\right) \right\} \frac{1}{E_0 T_0} d\tau \tag{5.6}$$

Therefore the viscous strain per cycle is

$$\varepsilon_y^p = \frac{\sigma}{2} \frac{\Delta t}{E_0 T_0} \tag{5.7}$$

5.3 Relationship between creep and relaxation

For a linear viscoelastic material, creep and relaxation are related by the convolution theorem using Laplace transformation. The time dependent stress is expressed

$$\sigma(t) = \int_0^t E(t-\tau) \frac{d\varepsilon(\tau)}{d\tau} d\tau \tag{5.8}$$

$$\varepsilon(t) = \int_0^t J(t-\tau) \frac{d\sigma(\tau)}{d\tau} d\tau \tag{5.9}$$

where $\varepsilon(t) =$ strain, $J(t)$ = the creep compliance and $E(t)$= the relaxation. Using s as the transform variable, $\sigma(s) = E(s)\varepsilon(s)$, and $\varepsilon(s) = J(s)\sigma(s)$. Therefore

$$E(s)\,J(s) = \frac{1}{s^2} \tag{5.10}$$

Taking the inverse Laplace transform, using the convolution theorem gives

$$\int_0^t J(t-\tau)E(\tau)d\tau = \int_0^t E(t-\tau)J(\tau)d\tau = t \tag{511}$$

Let the relaxation be of a power law type, such that

$$E(t) = Ct^n \tag{5.12}$$

The gamma function is defined by

$$\Gamma(x) = \int_0^\infty t^{x-1}e^{-t}dt \tag{5.13}$$

Using Eq. (5.12) and taking Laplace transform

$$J(t) = \frac{1}{C\Gamma(1-n)\Gamma(1+n)}t^n \tag{5.14}$$

54 Schapery's elastic-viscoelastic correspondence principle

Schapery (1984) proposed an extended elastic-viscoelastic correspondence principle applicable to linear or non-linear viscoelastic materials. It states that the constitutive equations for a viscoelastic material are the same for an elastic material, but with the actual stresses and strains replaced by pseudo stresses and strains. The constitutive equation for a linear viscoelastic material is represented by the following convolution integral

$$\sigma = \int_0^t E(t-\tau)\frac{\partial\varepsilon}{\partial\tau}d\tau \qquad (5.16)$$

wher σ = the stress, ε = strain and $E(t)$= relaxation modulus. Introducing the pseudo-strain

$$\varepsilon^R = \frac{1}{E_R} \qquad (5.17)$$

Eq. (5.17) may be written as $\quad \sigma = \frac{1}{E_R}\varepsilon^R \qquad (5.18)$

where σ, ε = physical stress and strain, respectively, E_R= an arbitrary constant with the same dimension as relaxation modulus (may be set equal to 1), and σ^R, ε^R = pseudo stress and strain, respectively. This formulation is usually simpler than other methods that use Laplace transforms. Schapery extended this simple method to obtain solutions to nonlinear viscoelastic boundary value problems with large deformations through the use of pseudo stress variables. For example, a viscoelastic constitutive equation using pseudo stress is given as

$$u_i = u_i^R \qquad (5.19)$$

$$\sigma_{ij} = \frac{1}{E_R}\int_{0-}^t E(t-\tau)\frac{d\sigma_{ij}}{d\tau}d\tau \qquad (5.20)$$

where u_{ij} and σ_{ij} = the physical displacement and the transpose of the nominal stress tensor, respectively, and u_i^R, σ_{ij} = pseudo displacements and the pseudo stress tensor, respectively.

5.5 Wnuk's viscoelastic crack growth theory

Wnuk (1979) proposed a nonlinear differential equation to describe the subcritical growth of a crack in a viscoelastic material as:

$$(G+M) \ \psi \left(\frac{\Delta}{\dfrac{dc}{dt}} \right) \tag{5.21}$$

where ψ = the normalized creep compliance function, Y = yield stress, and Δ is the inherent opening distance, assumed to be $\Delta = 0.125 \, K_{IC}^2 / Y^2$ (Ouchterlony, 1983) and M = the slow growth operator given by

$$M = 2Y \frac{d\sigma}{dc} \frac{\partial}{2\sigma} \frac{\Delta}{3} U_{tip} \tag{5.22}$$

where U_{tip} = $G/2Y$. The resulting equation for the fatigue crack propagation rate is per cycle

$$\frac{dc}{dN} = \frac{\pi}{768 \, Y^2} \frac{C}{f} \Delta K_I^2 \tag{5.23}$$

where $C = \dfrac{d\psi}{dt} \Big|_{t=0} \left(\dfrac{\Delta}{dc \, / \, dt} \right)$ and f = frequency of the applied load.

Example 6.1

Determine the rate of crack growth for a pavement in a parking lot, if the present length of the cracks comprise an 8 in. (20 cm) transverse and an 8 in. (20 cm) longitudinal crack that intersect, if the loaded truck wheels park in this area overnight for 8 hours daily. The pavement consists of 8 in.(20 cm) full-depth AC with dynamic compression modulus $E_1^* = $ 330,000 psi and a subgrade with modulus = 35,000 psi. The fracture toughness of the AC is K_{Ic} = 800 psi $\sqrt{in}$. The normalized creep compliance curve in bending from a Burger's model is described by parameters, $E_0 = 500000 \, \text{psi}$ $E_1 = 5000 \text{psi}, \ T_1 = 1900s, T_0 = 220s. \, (1 \text{ psi} = 6.895 \text{ kPa})$

Solution

For slow crack growth

$$E^* \rightarrow E^* / \psi(t)$$

where $\psi(t) = J(t)/J(0)$

$$J(t) = \frac{1}{E_0} + \frac{1}{E_1}\left(1 - \exp\left(-\frac{t}{T_1}\right)\right) + \frac{t}{E_0 T_0}$$

and

$$J(0) = \frac{1}{E_0}$$

Therefore

$$\psi(t) = 1 + \frac{E_0}{E_1}(1 - e^{-\frac{t}{T_1}}) + \frac{t}{T_0}$$

$$\frac{d\psi}{dt}\Big|_{t=0} = \frac{E_0}{E_1}\frac{1}{T_1} = 0.05$$

For $t = 8$ hours, or 28,800 s

The slow growth crack rate is

$$\frac{dc}{dN} = \frac{\pi}{768\,Y^2}\frac{C}{f}\Delta K_I^2$$

where $C = d\psi/dt\,|_{t=0} = 0.05$

For a through-crack

$$K_J = \sigma\frac{\sqrt{c\pi}}{1 + 0.125\lambda^2 c^2}$$

From **CHEVRON** the stress $\sigma = 40.7$ psi. $\lambda = 1.00$, c = 8.0 in. $K_J = = 22.5\,\mathrm{psi}\sqrt{\mathrm{in}}$, $Y = 302$ psi, so that

$$\frac{dc}{dN} = \frac{\pi}{768\,Y^2}\frac{0.05}{1/28800}K_I^2 = 0.43\ \text{in/cycle}$$

Therefore the crack growth rate is

$$\frac{dc}{dt} = 0.43\ in.\ (1.09\ \text{cm})/\text{day} \quad \leftarrow \textbf{Ans}$$

65

PROBLEMS

5.1

The creep-time curve for an asphalt concrete mix is given below.

Time, s.	0	5	50	500	5000	2000	5000	10000
Creep def., mm	0	10	30	50	75	100	200	300

From the creep curve deduce the corresponding relaxation curve and fit a Burgers model to the creep curve. Using the Burgers model calculate the relaxation curve and compare your results from the mechanical and the analytical model. If the applied stress and strain are 350 kPa and 0.075 mm/mm, respectively, determine the pseudo-strain, ε^R **u**sing the correspondence principle.

6

FATIGUE CRACK PROPAGATION

"The hallmark of a true theoretical solution is that it enables reliable predictions and integrity assessments from small-scale laboratory for large-scale structures with different geometry under different boundary conditions". Atkinson, 1980

Fatigue of viscoelastic/plastic materials

All highway materials suffer from fatigue cracking. The shape of the cracks resembles the *scales of the back of an alligator, hence the name "alligator cracking".* In the past the nature of fatigue was not understood. In 1958 the Navy Admiralty stated that they have not seen any fatigue cracking in ships. By 1965 they estimated only 15% of their ships had suffered from fatigue, but in 1970 when a ship broke in the middle because of fatigue cracking, they realized that the problem was much more severe than they estimated. Fatigue cracking is caused by cyclic loading which causes existing flaws in the material to grow to cracks that may lead to failure. The existence of flaws, whether arising from defects in the material, from fabrication or from service conditions, cannot be precluded from any material. Even diamonds contain inherent flaws. Such flaws can extend to cause catastrophic failure. Fracture mechanics circumvents the difficulty arising from the presence of sharp cracks in a stress analysis problem where there would be an infinite stress leading to fracture, by providing a parameter, the stress intensity factor, that characterizes the propensity for the crack to extend. The hallmark of a true fracture mechanics approach is that it should be possible to make reliable predictions and integrity assessments from small-scale laboratory structures under different boundary conditions. Even bold applications in composite technology to fibers having thicknesses comparable to the crack size are not atypical of fracture mechanics.

6.1 Mechanics of fatigue

Fatigue grows from inherent or starter cracks by blunting and resharpening (Frost, 1965) In the loading portion of a stress pulse a sharp crack in a tension field causes a large stress concentration at the tip which leads to a blunt or rounded crack tip. Plastic deformation has occurred in a small region embedded in elastic surroundings. During unloading the elastic region exerts compressive stresses on the plastically deformed region which close and resharpen the crack tip. This sequence of blunting and resharpening of the crack tip leads to fatigue crack propagation. A vast amount of experimental work has been done on fatigue crack growth over the last five decades. Paris (1967) was the first to introduce the stress intensity factor K in the study of fatigue of metals. His empirical law for the rate of crack propagation is

$$\frac{dc}{dN} = AK^n \qquad\qquad (6.1)$$

in which A and n are experimental constants. As a man of vision, Paris stated that we would have to develop "zillions of data points", using empirical methods and urged the profession to develop theoretical solutions to the problem.

6.2 Fatigue theory

Among the most notable works in this direction are those by Budiansky and Hutchinson (1978) and Kanninen and Atkinson (1980). The former work is based on Dugdale's model (1960) which is not fully representative of plane strain conditions for "long" cracks in metals typically found in engineering practice. Atkinson (1980) used a "superdislocation "model to derive a theoretical expression for the crack tip opening displacement (CTOD) in plane strain or small-scale yielding conditions.

Eq. (7.3) then simplifies to:

$$\delta_0 = 0.43 \frac{K^2}{EY} \tag{6.2}$$

According to Kanninen and Atkinson (1980) the symmetrical dislocation is an intrusion into the unbroken material at the crack tip. Therefore the net stretch of the plastic zone for an increment of ΔK is

$$\delta \bar{\Delta} = 0.153 \frac{(\Delta K - 0.47 \Delta K)^2}{EY} = 0.043 \frac{\Delta K^2}{EY} \tag{6.3}$$

The accuracy of this prediction for the crack opening displacement has been verified by a variety of both approximate and highly rigorous calculations of the CTOD Kanninen et al. (1977).

Using finite element analyses, Newman (1981) showed that the crack closure SIF is $K_{cl} = K_{op} \approx 0.25 K_{max}$. Kanninen and Atkinson's expression for the plane strain opening displacement during loading $\delta = 0.43 K^2 / (EY)$, means that the yield stress in plane strain is 2.32Y, where Y is the uniaxial yield stress in plane stress. Upon unloading the stress goes from 2.32Y to Y (Newman and Edwards 1988). Therefore the recoverable crack opening displacement from unloading is

$$\delta_r = \frac{0.43(0.70) K_{max}^2}{EY} \tag{6.4}$$

This is equivalent to a crack closure of $0.17 K_{max}$. Assuming that $K_{cl} = K_{op}$, during cyclic loading, the total value of $K_{op} = 0.47 K_{op} (R = 0)$. The resulting expression for the net stretch of the process zone is

$$\delta \bar{\Delta} = 0.153 \frac{(\Delta K - 0.47 \Delta K)^2}{EY} \, 0.043 \frac{\Delta K^2}{EY} \tag{6.5}$$

$$\delta \bar{\Delta} = 0.043 \frac{\Delta K^2}{EY} \tag{6.6}$$

This is a key relationship for FCGR in metals for plane strain.

6.3 Extension of Griffith's theory of fracture to include fatigue

Fatigue is considered as a sequence of crack growth steps under cyclic loading. There is a tiny zone at the tip of the crack called the process zone, where there is intense activity during fatigue crack growth. According to Rice and Johnson (1970), the process zone size $\bar{\Delta}$ is a controlling factor for quasi-static extension of a crack in an elastoplastic solid. This concept plats a major role in the proposed analytical model for the rate of fatigue crack propagation. Ramsamooj (1999) hypothesized that a crack cannot extend unless preceded by the extension of the process zone. The energy balance must be satisfied during the entire growth period, so that the sum of the strain energy release must be equal to the energy demand to create the new surfaces. As the process zone is a region of intense activity, the strain energy release rate there must be critical or equal to G_c. Accordingly

$$G \delta_c + G_c \delta \bar{\Delta} = G_c \delta_c \tag{6.7}$$

The first term in Eq. (6.8) is Griffith's elastic strain energy release rate and the second term is the strain energy rate due to the increment in the process zone, a "new term" which extends the Griffith criterion for fracture to include fatigue.

From Eq. (6.8) the rate of crack growth per cycle is

$$\left(\delta_c \right)_{cycle} = \delta \bar{\Delta} \frac{G_c}{G_c - G} \tag{6.8}$$

where $G K^2 (1 - \upsilon^2) / E$. Therefore the FCGR for Mode I is

$$\frac{dc}{dN} = 0.043 \frac{\Delta K^2}{EY\left(1 - K_n^2\right)} \tag{6.9}$$

where $K_n^2 = \left(K_{max} / K_{Ic}\right)^2$. This is the final expression for fatigue crack propagation rate in *metals*.

6.4 Fatigue threshold for metals

Fatigue tests are very time consuming to determine in the laboratory. In order to reduce the time for experiments on specimens without notches, a threshold value of the SIF was introduced. The threshold ΔK_{th} at which the FCGR is according to the British Standards is 10^{-11} mm/cycle whereas according to ASTM it m/cycle. Typically the difference between the two fatigue lives using the threshold as that for zero crack growth is 14-38% (Taylor, 1988). For the great majority of fatigue crack growth rate data, the British value falls within 10% of the true asymptote to zero crack growth rate. From the designer's point of view, if only one is given without a statement of definition, it is prudent to use the conservative value. Marci (1992) found that a fatigue threshold has to be entirely above K_{op}, and has to be the upper bound amplitude of cycles for a given K_{max}, the action of which does not produce FCG irrespective of how these amplitudes are applied. When $\Delta K = \Delta K_{th} = K_{max} - K_{min}$

$$\frac{dc}{dN} = 0.043 \frac{\Delta K_{th}^{\;2}}{EY\left(1 - K_n^2\right)} = 10^{-10} \, \text{m/cycle} \tag{6.10}$$

Let $R = K_{min} / K_{max}$. At the threshold $\Delta K_{th} = K_{max}\left(1 - R\right)$

$$\frac{dc}{dN} = 0.043 \frac{\Delta K_{th}^{\;2}}{EY\left(1 - K_n^2\right)} = 10^{-10} \, \text{m/cycle} \tag{6.11}$$

For a constant value of K_{max}

$$\frac{dc}{dN} = 0.043 \frac{(K_{max}(1-R) - \Delta K_{th})^2}{EY(1-K_n^2)} = 10^{-10} \, \text{m/cycle} \tag{6.12}$$

For the same K_{max}, for $= 0$

$$\frac{dc}{dN} = 0.043 \frac{(K_{max} - \Delta K_{th,0})^2}{EY(1-K_n^2)} = 10^{-10} \, \text{m/cycle} \tag{6.13}$$

Comparing the last two equations

$$\Delta K_{th,0} = \Delta K_{th}(1-R) \tag{6.14}$$

This is the general relationship for the variation of the fatigue threshold with the stress ratio R and constant value of K_{max}. It is well corroborated by the experimental data of Marci (1992). Schmidt and Paris (1973) found experimentally that there is a limiting value of R, because of crack tip blunting. They also derived a theoretical expression for R at closure. When the material has a threshold SIF or ΔK_{th}, the complete expression for the rate of crack propagation is

$$\frac{dc}{dN} = 0.043 \frac{1}{EY(1-K_n^2)} (\Delta K - \Delta K_{th})^2 \tag{6.15}$$

where $K_n = \left(\dfrac{K_{max}}{K_{Ic}}\right)^2$.

For combined loading in Modes I and II, the equivalent Mode I loading is (Tian,1997)

$$K_{Ieq} = \sqrt{K_I^2 + 0.866 K_{II}^2} \tag{6.16}$$

For combined loading in Modes I, II and III, the equivalent Mode I loading is

$$K_{Ieq} = \sqrt{K_I^2 + 0.866\,K_{II}^2 + K_{III}^2}$$ (6.17)

6.5 Physical meaning of parameters

All materials contain inherent flaws, and the most common natural flaw shape is semi-elliptical with an aspect ratio of 1 to 4 (vertical/horizontal) as is assumed in this books. The initial crack length or the size of the inherent defect c_0 is obtained from Ouchterlony (1983) as:

$$c_0 = 0.05\frac{K_{Ic}^2}{f_b}$$ (6.18)

where K_{Ic} is the fracture toughness, determined experimentally, and f_b = the bending strength, determined experimentally from tests on beams. The threshold SIF ΔK_{th} is metals has not been predicted. It has been determined experimentally for most metals, and it is treated as a constant material property. However, as pointed out by Marci, it is dependent on the value of K_{max}. Once it is known for one ratio R it can be predicted for other values of R.

6.6 Fatigue of composites

The principles of linear fracture mechanics (LEFM) are applicable to composite materials (Corten, 1972). For any elastoplastic the foregoing fatigue theory should be applicable. However, there is no experimental evidence to indicate that the full closure mechanisms discussed above for metals is applicable. Accordingly the fatigue life of metals can be modified for unreinforced composites as:

$$\frac{dc}{dN} = 0.153\frac{1}{EY\left(1-K_n^2\right)}\left(\Delta K - \Delta K_{th}\right)^2$$ (6.19)

There is unfortunately a dearth of published data giving the material properties, E, Y, K_{Ic} and . The available data for Polycarbonate, Nylon 6.6 and Polymethylmethacrylate (PMMA) have been used to validate the FCGR of these materials.

6.7 Validation of the theory of fatigue for metals using published data

A substantial amount of published data is presented to validate the preceding fatigue theory for predicting the fatigue crack growth rate for steel, aluminum, titanium, and nickel alloys.

The test data include all of the published data in which the material properties were included. The material properties, type of test, thickness of specimens, etc. are presented in Table 6.1. The comparisons for the theoretical predictions for some of the metal alloys are presented in Fig. 6.1 through Fig. 6.2, and for some of the composites are presented in Fig. 6.3. There is good agreement between the theory and the experimental data, indicating that the fracture mechanics model can be used to predict the fatigue life of metal alloys as well as unreinforced composites.

Table 6.1 Mechanical properties of metals and metal alloys

Metal	E (MPa)	Y(MPa)	$K_{Ic}\left(MPa\sqrt{m}\right)$	$R/\Delta K_{th}\left(-MPa\sqrt{m}\right)$	Th./Orient.(-/mm)
Steel alloys					
A533B-1	207^a	483^b	220^b	$0.1/8^b$	CT/5 RW
A508-2	207^a	483^b	220^b	$0.1/8^b$	CT/5 RW
A517—Gr. F	207^a	758^c	151^d	$0/7.0^d$	TR^e
A36	207^a	248^d	151^d	$0/11^d$	
ABS-C	207^a	269^d	151^d	$0/12.1^d$	
304/316	207^a	483^b	220^b	$0.3/6.46^f$	CT/30
4340	207^a	862^a	99^a	$0.3/6.46^f$	
D6ac	207^a	1641^g	38.5^g	$0.3/3.0^g$	CT/25
D6 II 58	207^a	655^j	38.5^g	$0.3/3.0^g$	CT/25
D6ac (C)	207^a	1007^f	38.5^g	$0.3/3.0^g$	CT/46
HY-80	207^a	1007^f	38.5^g	$0.3/3.0^g$	CT/50
HY-130	207^a	1007^f	38.5^g	$0.1/4.5^g$	CT/50

10Ni-Cr-Mo-Co	207^a	1269^j	155^j	$0 / 2.85^j$	CT/50
12Ni-5Cr-3Mo	207^a	1269^j	155^j	$0 / 2.45^j$	CT/50
8630	207^a	985^j	104^i	$0.5 / 4.1^j$	CT/8.0
Aluminum alloys					
7076-T6	68.9^a	540^g	30^g	$0.05 / 1.7^g$	L – T
7076-T6	68.9^a	540^g	70^g	$0.05 / 1.7^g$	L – T
7076-T3651	68.9^a	441	35.2^g	$0 / 2.5^g$	
7076-T6	68.9^a	540^g	70^g	$0.05 / 1.7^g$	L – T
2024-T3	68.9^a	345^g	44^m	$0.0 / 1.87^k$	L – T
2024-T3	68.9^a	393^f	120.7^f	$0.0 / 5.5^f$	C – T
5456H321	68.9^a	255^k	65^k	$0.0 / 1.87^k$	
7075-T6	68.9^a	531^d	319^k	$0.0 / 2.0^k$	DCB
7079-T6	68.9^a	225^d	65^h	$0.0 / 1.89^k$	
2219-T651	68.9^a	345^n	65^h	$0.0 / 1.8^n$	*DCB*
2219-T851	68.9^a	345^n	76.7^n	$0.0 / 1.8^n$	CT
Titanium alloys					
Ti-6Al-4V	116^o	1041^j	$81,5^j$	$0 / 6.6^k$	
RA-Ti-6Al-4V	116°	834^p	97.4^p	$0.5 / 4.84^p$	CT
Nickel alloys					
Inconel 718	200^a	786^a	111^a	$0 / 7.12^q$	
Inconel 750	214^a	1103^a	142.3^a	$0 / 7.12^q$	

a Hertzberg (1995)

b Paris (1971)

c Bucci et al. (1971)

d Barsom and Rolfe (1987)

e Balladon et al. (1980)

f McEvily (1977)

g Mautz and Weiss (1976)

j Barsom et al. (1971)

k Meguid (1989)

l Stephens (1986)

m De Koning (1981)

n Chang et al. (1981)

o Barsom and McNicol (1974)

p Katcher and Kaplan

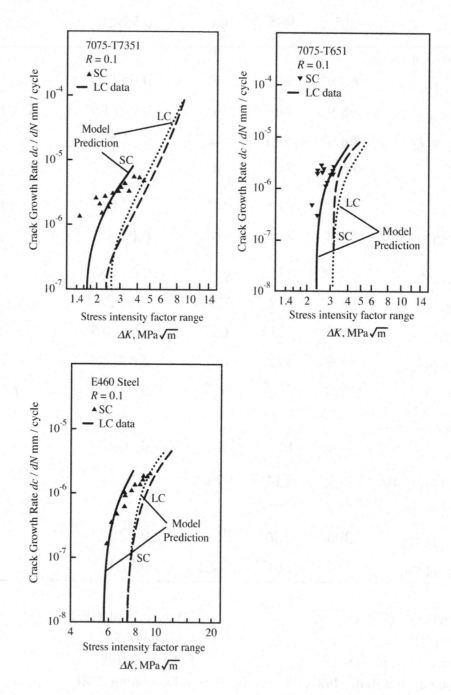

Figure 6.1. Crack Growth Rate as a function of *DK* for Aluminum and Steel

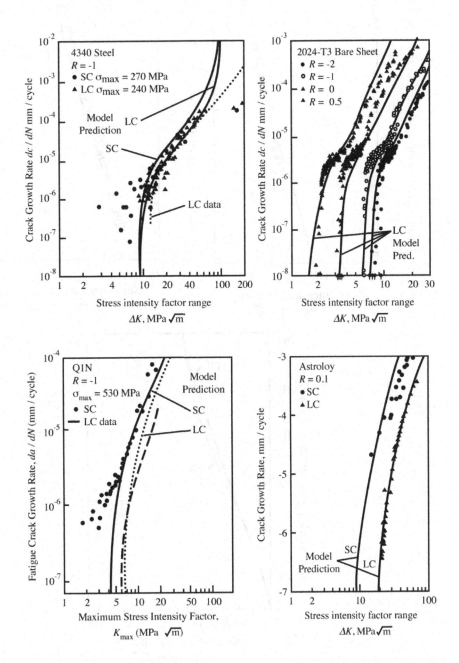

Figure 6.2. Crack Growth Rate as a function of *DK* for Steel and Astroloy

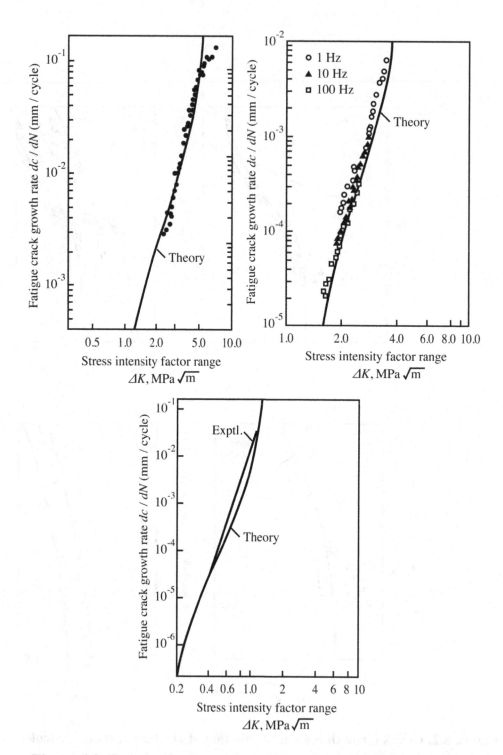

Figure 6.3. Fatigue of composites as a function of *DK* for Composites

7

FATIGUE IN CONCRETE, ASPHALT CONCRETE, PLASTICS, AND STABILIZED SOILS

The theoretical framework for the applicability of fracture mechanics to metals was developed for small amounts of the plastic deformation at the crack tip, such that the crack size and the nearest boundary to the crack tip should be at least at least 25 times the size of the plastic zone (ASTM E 399-83). In asphalt concrete or concrete this is easily satisfied because of the large sizes of the inherent defects and the relatively low values of the stress intensity factors for typical pavements. Taylor (1989) suggested that for LEFM or continuum mechanics may be applied to metals when the size of the crack is approximately ten times larger than the average grain size or other relevant microstructural distance. For concrete or asphalt concrete with a maximum size aggregate of 25 mm, the representative size taken as the 15% size is about 0.6 mm, so that the continuum mechanics is not strictly applicable for cracks smaller than 6 mm. However, as shown later, the starter crack or the inherent defect in PCC or AC is approximately this size, so that this guideline is met. After a crack is loaded and then unloaded the crack tip is not fully open upon reloading, because of plasticity induced closure and closure from asperities, etc. so that the net $K_{op} = 0.47K_{max}$, as discussed in Section 6. However for asphalt concrete, concrete and polymers, this may not be true. There has been no research on this topic, so that it is prudent to neglect the effect of closure.

7.1 Analytical Expression for the fatigue life of geologic materials

The crack opening displacement for an elastoplastic material for plane strain in the *opening mode* is inferred from the Dugdale model as

$$v \approx \frac{8}{9\pi^2} \frac{K_1^2}{EY} \tag{7.1}$$

For a viscoelastic/plastic material such as asphalt concrete or polymer concrete, the crack tip opening displacement, or the crack width, for plane strain in the opening mode is

$$v \approx \frac{8}{9\pi^2} \frac{K_1^2}{EY} \psi(t) \tag{7.2}$$

Because highway fatigue is caused mostly by bending, the terms EY will be replaced by $E_b^* f_b$ for viscoelastic/plastic materials in bending. The magnitude of the complex modulus $E^* = E/\psi(t)$, where $\psi(t)$ is the normalized creep compliance function, so that the crack tip opening displacement becomes

$$v \approx \frac{8}{9\pi^2} \frac{K_1^2}{E^* Y} \tag{7.3}$$

in which E^* is the *magnitude* of the dynamic modulus and Y is the tensile strength f_b of the asphalt concrete. Therefore the complete general expression for the crack growth per cycle in concrete (PCC) or asphalt concrete (AC) is

$$\frac{dc}{dN} = 0.153 \frac{1}{EY\left(1 - K_n^2\right)} \left(\Delta K - \Delta K_{th}\right)^2 \tag{7.4}$$

$E^* =$ dynamic modulus in *bending*, $f_b =$ the tensile *bending* strength and $K_{Ic} =$ the fracture toughness.

7.2 Effect of thermal stresses on fatigue of geologic materials

Eq.(7.4) governs the subcritical growth of the crack up to the point of instability in which N = number of applied cycles, c = crack length, $E^*(t, T)$ = magnitude of the complex modulus, a function of time and temperature, $f_b(T)$, $K^t(T)$ = the bending strength and fracture toughness respectively, are functions of temperature; $K(\sigma,T)$, and $\Delta K(\sigma,T)$ = the applied SIF and its increment, respectively, are functions of the applied stress σ and temperature T.

When the temperature of a concrete slab decreases with depth then the greater thermal expansion at the surface causes warping so that the central area is higher than the edges. Similarly the when the temperature increases with depth the slab warps upwards at the edges. Because the vertical movement is restrained by the weight of the slab, warping stresses are created in the slab. For a uniform temperature gradient, with the top warmer than the bottom, the restrained tensile warping stress in the undersurface of the slab is

$$f_{rw} = \frac{E\alpha T}{2(1-v)} \qquad (7.5)$$

For a uniform thermal expansion restrained by the friction of the subgrade, the restrained tensile stress at the undersurface is

$$\sigma_{rt} = \frac{E\alpha T}{1-v} \qquad (7.6)$$

The combined effect of vehicular, thermal axial and warping stresses can be easily obtained. Croney recommended a "safe" value of 0.8 MPa to be added to the vehicular stress. Thus the thermal stresses may contribute to initiating a crack, but in the fracture mechanics method of computing fatigue, the thermal stresses merely serve to change the values of K_{max} and K_{min}, so that the value of $\Delta K = K_{max} - K_{min}$ remains the same, but the value of K_n changes. Since the thermal cyclic loading occurs only once per day, the fatigue effect of thermal stresses is relatively insignificant.

7.3 Validation of the fatigue life of concrete using published data

The size of the starter defect, the fracture toughness and the rate of crack propagation are given by the same formulae for concrete, asphalt concrete, and EVAPAVE as stated above. The modulus of rupture or the bending strength of concrete is $f_r = 10\sqrt{f_c'}$. The compressive modulus is least six times stronger than the tensile modulus of concrete. The bending modulus is given by

$$E_b = E_t \left(\frac{2}{1 + \sqrt{\dfrac{E_t}{E_c}}} \right) \tag{7.7}$$

where E_c is the compression modulus of concrete, usually taken as

$$E_c = 57000 \sqrt{f_c'} \,.\text{psi} \tag{7.8}$$

From Eq. (7.7) the **bending** modulus of concrete is $0.24 E_c$

According to Nordby (1958) PCC has an endurance limit of 0.52-0.55 f_r where f_r is the modulus of rupture. This was contradicted by Kesler (1953) who found no endurance limit within 10 million cycles. However as explained in Section 6, there is an endurance limit, but it has not yet been determined conclusively. It may be deterministic, after the problem of the rate of crack propagation of *very short cracks* (Newman, 1998) has been solved. Experimental evidence indicates that for asphalt concrete for the low strain levels in high cycle fatigue of highway and airport pavements, the endurance limit is approximately

$$f_{el} = 60 \left(10^{-6} \right) E_b \tag{7.9}$$

7.4 The *J*-Integral at high stress levels

Concrete can perform at relatively higher stresses than AC. At these higher stresses, it is more appropriate to use the *J*-integral instead of the SIF. J_R is path independent, providing that the integration path is outside the inelastic zone. It is based on the deformation theory in plasticity which is strictly only applicable to a stationary crack. Hutchinson and Paris (1973) formulated the following guidelines for the applicability of J_R in crack propagation studies:

$$\frac{J_R}{dc} = \frac{24 J_{Ic}}{\Delta c} \quad b > 40 \frac{J_{Ic}}{\Delta c} \tag{7.10}$$

Where b = the size of the ligament ahead of the crack. The associated crack extension must be of the order of a few percent of the remaining ligament. The ratio of the values of J_R for linear and nonlinear conditions (Rice 1968) is

$$\frac{J}{J_{lin}} = \lambda \tag{7.11}$$

Kanninen and Popelar (1985) give the value of the dimensionless parameter as

$$\lambda = \frac{8}{\pi^2} \left(\frac{f_b}{f_r} \right)^2 \ln \left\{ \sec \left(\frac{\pi f_r}{2 f_b} \right) \right\} b \tag{7.12}$$

where f_b and f_r = the bending stress and the modulus of rupture, respectively.

$$J_{lin} = \frac{K^2}{E} \left(1 - v^2 \right) \tag{7.13}$$

For small values of f_b / f_r the ratio of $\lambda \to 1$, but as $f_b \to f_r$, $\lambda \to \infty$. Therefore **Eq.(7.4)** for fatigue can be modified for both high and low cycle fatigue by simply replacing ΔK^2 by $\lambda \Delta K^2$.

The published data from several researchers are listed below, together with their experimental data. The model predictions based on the supporting equations is made by a computer program *ACBEAMF*.

7.5 Validation of the analytical model predictions for concrete

The preceding fatigue theory is applicable to concrete and other elastoplastic materials providing that LEFM is satisfied. For very large stresses approaching ultimate strength collapse, the failure criterion is the modulus of rupture calculated on the net remaining section. For intermediate stresses, there is interaction between fatigue and ultimate strength collapse given

$$\frac{K_{\mathrm{I}}}{K_{\mathrm{Ic}}} + \frac{f_b}{f_{\mathrm{r}}} = 1 \qquad\qquad (7.14)$$

For typical concrete $\dfrac{f_b}{f_{\mathrm{r}}} \approx 0.8$.

7.5.1 Murdoch's data (1953)

A very comprehensive investigation consisting of 54 flexural fatigue tests on plain concrete beams was conducted on four-point bend specimens (FPB). The speed of testing varied from 70-440 cycles per minute and the flexural stress from 44-91% of the modulus of rupture. One hundred specimens were tested in flexure on the broken parts of the beams after fatigue failure to minimize the variances from specimen to specimen. The details of the mix composition, compressive strengths, and specimen geometry used are as follows:

Table 7.1 Properties of test specimens

Mix Composition	Cement Aggregate	Type I Portland Cement well-graded Wabush river sand and gravel, 25.4 mm max. size agg.
Compressive strength at 90 days		24,800-31,700 kPa
Modulus of rupture		2,810-3,870
Age was tested		90 days
Specimens: Beams		15.24 x 15.24 x 152.4 cm
Cylinders		15.24 x 30.48 cm
Min/max cyclic stress R		0.07
Type of test		FPB Electrohydraulic pulses
Frequency of test		1.2-7.3 Hz

The graph of the test data versus the stress ratio (bending stress/modulus of rupture) for three frequencies of loading is presented graphically in Fig. 7.1-7.2. Evidently the frequency of loading has little effect on the fatigue life.

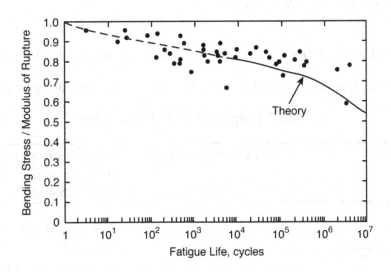

Figure 7.1. Fatigue life of concrete (Kesler, 1953).

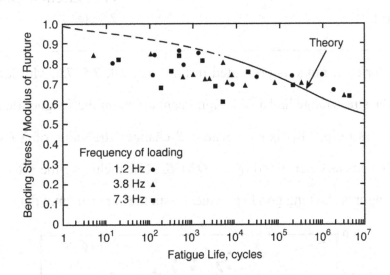

Figure 7.2. Fatigue life of concrete (Kesler, 1953).

7.5.2 Murdoch and Kesler's data (1959)

One hundred and seventy-five specimens were tested in cyclic loading without stress reversal. The cyclic stress ratio varied from 0.13 to 0.75. The details are given below

Table 7.2 Properties of test specimens

Mix Composition	Cement Aggregate	Type I Portland Cement well-graded Wabush river sand and gravel, 25.4 mm max. size agg.
Slump		25.4 – 152.4 cm
Compressive strength at 90 days		38,610 kPa
Specimens: Beams		15.24 x 15.24 x 152.4 cm
Cylinders		15.24 x 30.48 cm
Min/max stress, R		-0.13-0.25; 0.25; 0.50; 0.75
Type of test		FPB. Electrohydraulic pulses
Frequency of test		7.5 Hz

The stress data for each range is presented in Figs. 7.3, 7.4, 7.5, 7.6. Murdoch and Kesler concluded that the stress range had a very significant effect on the fatigue life of a specimen at a specified stress ratio. This is so, because R changes the value of ΔK, and ΔK. The theoretical predictions as determined by **FCONCBEAM**, including the effect of R are also presented in the figures showing good agreement with the experimental data.

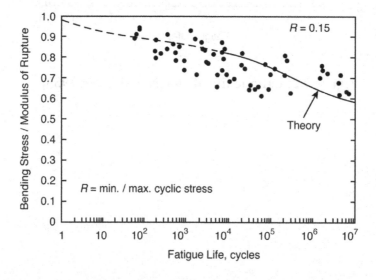

Figure 7.3. Fatigue life of concrete (Murdoch and Kesler, 1959).

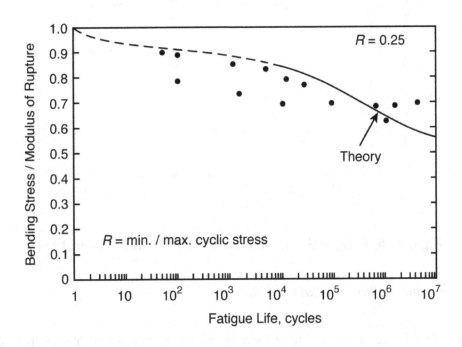

Figure 7.4. Fatigue life of concrete (Murdoch and Kesler, 1959).

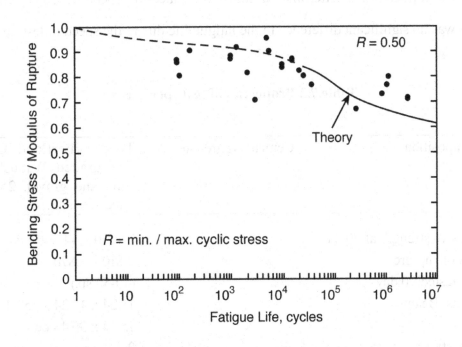

Figure 7.5. Fatigue life of concrete (Murdoch and Kesler, 1959).

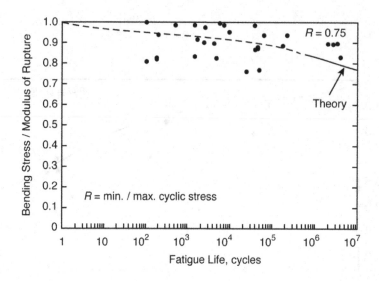

Figure 7.6. Fatigue life of concrete (Murdoch and Kesler, 1959).

7.5.3 Raithby and Galloway's data (1974)

The effects of moisture, age and rate of loading on the fatigue life of plain concrete beams were investigated on three types of concrete, PQ1, PQ2, LC1. It was found that oven-dried specimens performed much better than saturated or surface-dry specimens. They also found that there was no significant difference in the fatigue life due to the speed of testing.

Table 7.3 Properties of test specimens

Mix Composition	Cement Aggregate	Type I Portland Cement well-graded Wabush river sand and gravel, 25.4 mm max. size agg.
Compressive strength at 90 days		21,000 – 44,800 kPa
Modulus of rupture		2,810 – 3,870
Age was tested 90 days		(40-50 sp.)
Specimens: Beams		15.24 x 15.24 x 152.4 cm
Cylinders		15.24 x 30.48 cm
Min/max stress		0
Frequency of test		20 Hz

analytical predictions by **FCONBEAM** are presented in Figure 7.7. The agreement is good.

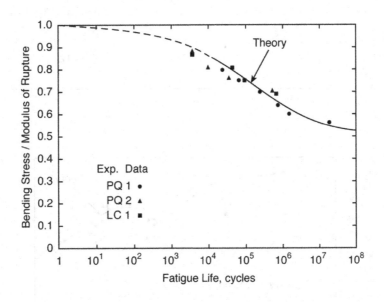

Figure 7.7. Fatigue life of concrete (Kesler, 1953).

7.5.4 Ballinger's data (1973).

Fatigue tests were conducted on specimens moist-cured for 7 days, then stored in the laboratory for 425 days. Cyclic fatigue tests were conducted at a frequency of 7.5 Hz, with stress ratio $R = 0.15$. The test data on fully saturated specimens for three types of concrete, together with the analytical predictions by **FCONBEAM** are presented in Fig. 7.8. The agreement is good.

Table 7.4 Properties of test specimens

Mix Composition	Cement Aggregate	Type I Portland Cement well-graded Wabush river sand and gravel, 25.4 mm max. size agg.
Compressive strength at 90 days		42,990 kPa
Flexural strength		4300 kPa
Modulus of rupture		2,810-3,870 kPa

Age was tested 90 days	(44 sp.)
Specimens: Beams	15.24 x 15.24 x 152.4 cm
Cylinders	15.24 x 30.48 cm
Min/max stress	0.15
Frequency of test	7.5 Hz

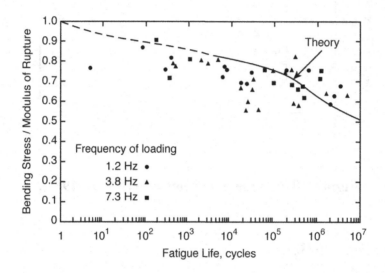

Figure 7.8. Fatigue life of concrete (Ballinger, 1974).

7.6 The scatter of the results

Concrete contains an abundance of flaws-holes, tiny cracks, pre-cracked aggregates, coalescence of flaws, segregation, lack of bond, etc. The large variation in the magnitude and distribution of these flaws leads to large scatter in the rates of crack growth and therefore in the fatigue life. Considering the nature of fatigue tests on concrete and the inherent large scatter of results, the overall agreement between the theory and the experiments for a total of 364 specimens is good. The practical significance of this is that the fatigue life of concrete can be predicted from simple tests, using the material properties with the appropriate loading and boundary conditions.

7.7 Validation of the analytical prediction of fatigue life of asphalt concrete

7.7.1 Majidzadeh et al.'s data (1976)

A substantial amount of published fatigue data for a large number of asphalt concrete mixes is used to validate the new analytical expression for crack growth rate. The raw test data was extracted from Majidzadeh et al. (1976), include dynamic modulus, indirect tensile strength, fracture toughness, creep and fatigue life of asphalt concrete beams consisting of three types of asphalt cement with numerous additives, three asphalt contents, three filler asphalt ratios and several gradations

However, only one creep compliance test was done for 54 test mixtures, leaving room for large scatter of results. The raw test data included dynamic modulus, ultimate tensile strength, critical stress-intensity-factor and fatigue life of asphalt concrete beams fabricated several types of asphalt cements. Five asphalt contents, three filler asphalt ratios, six additives and several gradations as shown in Table 7.5. Over 100 fatigue tests were conducted using 2" x 2" beams on an elastic foundation with modulus of 250 psi. The loads vary from 45 lb. to 200 lb. using a haversine with 5 Hz interrupted to give 2 cycles per second. The values of the compression dynamic modulus E^*, the critical stress intensity-factor KIc, and the indirect tensile strength σ_{yi} for each of the mixtures tested is presented in Table 7.6.

The dynamic moduli were measured at low stress levels where the values are dependent on the frequency of loading, which was kept constant in both the dynamic moduli and the fatigue tests. The bending strength $f_b \approx 1.5 f_t$, and the $f_t = 2/3\sigma_{yi}$, so that The dynamic moduli of the asphalt concrete mixes were determined using cylindrical compression tests.

Table 7.5 Details of AC mixtures from Majidzadeh et al (1976)

AC mixture (pen. grade)	Asphalt % by wt.	A Mix 4.5% Filler	B Mix 9.0% Filler	C Mix 13.5% filler
X (60/70 pen.)	5-9	x	x	x
Y (85100 pen.)	5-9	x	x	x
Z (120/150 pen.	5-9	x	x	x

tests, but the fatigue tests used beam on elastic foundation tests. It is well known that the tensile modulus of asphalt concrete is much greater than the compressive modulus. The ratio of the initial compression/tensile moduli as well as the corresponding creep rates were about 4. Accordingly, Eq. (7.7) shows the value of the bending modulus $E_b = 0.33 \ E_c$. The computer program *ACBEAMF* determined the analytical values of the fatigue life of each beam tested. The stress ratio, defined as the bending stress/ultimate bending strength f_b / f_{ct}, for the AX, AY and AZ mixes as a function of the experimental fatigue and the theoretical "predictions" are presented graphically in Figure 7.9 for the A mixtures, Figure 7.10 for the B mixtures and Figure 7.11 for the C mixtures. There were also fifteen tests with several additives. These results are presented graphically in Figure 7.12. The agreement between the predictions of the theory and the experimental data is good.

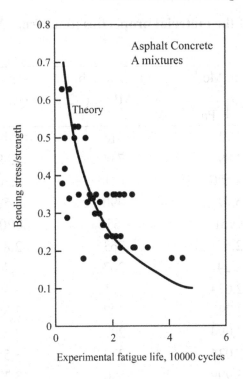

Figure 7.9 Comparison of theoretical predictions and the experimental data for the A mixtures (raw exptl. data from Majidzadeh et al. (1976)

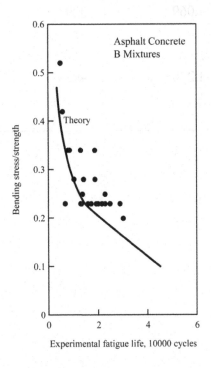

Figure 7.10 Comparison of theoretical predictions and the experimental data for the B mixtures (raw exptl. data from Majidzadeh et al. (1976)

TABLE 7.6. Summary of the material properties (data from Majidzadeh et al. (1976)

Asphalt Concrete	Dyn. Mod $\lvert E^* \rvert$ kPa	Frac. Tough. KIc, MPa-m1/2	Ten. Str. syi, kPa	Fat. Life Nf (av.) cycles
AY-7	1,724	418	3,448	25,000
A/B2922/7	1,896	734	4,895	30,000
A/B3014/7	4,550	631	3,300	40,000
A/B3603/7	5,792	990	7,585	90,000
AY-5	1,951*	265*	2,206*	6,300
Latex-170 (L3)	2,551	263	2,841	25,500
Latex-170 (L5)	2,241	207	2,765	46,200
LVP-100 (T3)	2,034	221	3,468	57,000
LVP-100 (T5)	1,793	207	3,551	72,900
Sulfur (S1-9)	2,094	297	3,310	27,700
Sulfur (S1-3)	3,034	330	3,551	38,500
Sulfur (S1-1)	3,999	550	3,723	61,500
AX-6	2275	292	2655	14,600
AY-6				36,800
AZ-6				10,900
BX-6	1,069	429	2,517	19,300
BY-6				10,200
BZ-6				16,900
CX-6	1,276	495	4,344	36,800
CY-6				9,300
CZ-6				27,800

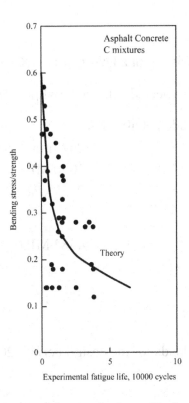

Figure 711 Comparison of theoretical predictions and the experimental data for the C mixtures (raw exptl. data from Majidzadeh et al. (1976)

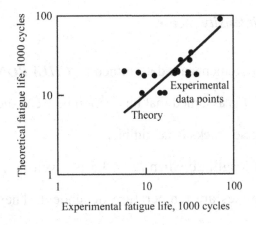

Figure 712 Comparison of theoretical predictions and the experimental data for AC + admixtures (raw exptl. data from Majidzadeh et al. (1976)

7.8 Validation of the analytical prediction of the fatigue life of full-scale flexible pavement

A typical full-depth pavement consisting of a DGAC mix AX-6 surface course on a silty clay subgrade as shown in Figure 1.4 is selected. The material properties are:

Surface: AX-6 (Data from Majidzadeh et al., 1976)

Thickness	244 mm
Dynamic modulus (bending)	1000 MPa
Flexural tensile strength	2655 kPa

Base Course

Dynamic modulus	4482 MPa

Subgrade: Gravel

Dynamic modulus	276 MPa

Traffic

No. of equivalent 80 kN-axle loads for the design period, 20 years   20 million

Temperature

Sinusoidal variation $0 - 45^o\,C$

7.4.1 Stresses and stress intensity factors

The stresses in the flexible pavements are determined by ***CHEVRON***. The starter cracks are initiated in the wheel track. It is assumed that interaction between the dominant crack tips in the longitudinal and transverse cracks is negligible.

The SIF for a crack of length $2c$ when $\lambda c < 3.5$ is given by Eq. (4.11). The material properties of the DGAC mixes are temperature dependent. They are approximated by (Majidzadeh et al., 1976):

Dynamic modulus: $E^* = E^*_{25^o C}\, e^{-0.0315(T-25)}$ $\qquad\qquad$ (7.15)

where T = temperature in degrees C.

Indirect tensile strength: $f_{ct=} \; f_{ct\,25^\circ C} \, e^{-0.0173(T-25)}$ (7.16)

The bending stress determined from **CHEVRON** is also temperature dependent, since the dynamic moduli are functions of the temperature.

7.9 The Endurance limit

ASTM standard defines the endurance limit in cyclic loading as the stress level at which the crack growth rate is 10^{-7} *mm/cycle, whereas the British standard defines the crack growth rate as* 10^{-8} *mm/cycle.* Typically the difference between the two values range from 14 to 38% (Taylor, 1988). For the great majority of data, an endurance limit of 10^{-8} mm/cycle fall within 10% of the true asymptote to zero crack growth. *This facilitates the computation of fatigue life for typical pavements, which is accordingly defined as the number of cycles for the crack to grow from the starter crack to the crack length at failure.*

The endurance limit is not fully understood. Even the very existence of the endurance limit is questioned. In the published fracture mechanics research, a threshold value of the stress intensity factor has been determined experimentally for many metals. There is no analytical method for determining the value of ΔK_{th}, because the mechanism of very small crack growth is not known at the present time. There is an excellent discussion of the state-of-the-art of very small crack growth by **Newman (1998)**

From experimental evidence, it seems reasonable to use an endurance limit for all aggregative materials such as asphalt concrete, cement concrete, composites and EVAPAVE. **However, the micro strain level varies as shown in Table ??**

7.10 Computer program *ALLIGATOR* (App. V)

This program was used to predict the fatigue life of the full-scale AC pavement. It computes the fatigue life as 5.06 million cycles in 20 years. In The AASHO Road Test a similar AC pavement was tested. Comparison of the predicted fatigue life of the DGAC mix in the full-scale pavement and that in The AASHO road test shows that the "prediction" is reasonable.

The Results

The importance of using fundamental material properties and a realistic performance model that can describe the behavior of the pavement in terms of the boundary and geometric conditions cannot be over-emphasized. The results show that the fatigue life in the field is quite different from that of the laboratory beam tests. In general, the fatigue lives in the field are much greater than those obtained from laboratory fatigue tests, using the same stress level as in the field. The major reason for this large difference is that the stress-intensity-factor is inversely proportional to fourth root of the coefficient of subgrade reaction for the *same bending stress*. Therefore laboratory tests in which the only variable is bending stress cannot directly predict full-scale performance with any degree of reliability.

7.11 Top-down fatigue cracking

"AC top-down cracking appears to be not thoroughly understood and is generally not considered as a causative factor for pavement cracking" Mahoney (2011), but Florida DOT reports that it is the dominant type of cracking for their AC pavements due for rehabilitation. The period of initiation is about 3 to 8 years. A brief summary of various reports of top-down cracking from some countries was presented by Mahoney as follows.

Top-down cracks in AC (**TDC**) pavements form within 3 – 8 years in many countries:

France: 3-5 years.

UK: Within 10 years for pavements 180 mm or thicker.

Netherlands: 160 mm. or greater.

Japan: 1-5 years.

California: analysis showed truck tire edges causing high surface strains.

Washington State: 3-8 years with thicknesses of 160 mm or more.

Florida: 5-10 years with wide range of thicknesses.

There are three basic views of causative factors:

    (1) High surface horizontal tensile stress,

    (2) Hardening of the binder from aging of the AC, and

    (3) Low stiffness in the surface layer caused by high temperatures.

Some experts support the view that the TDC could form due to truck tire edge stresses that produce high surface tensile strains. *Immediately below the tire*, the stress analysis from **CHEVRON** shows that the AC is under a compression stress of the order of magnitude of the tire pressure. The stresses that can counteract this stress are from braking and to a lesser extent driving traction and from thermal changes. Severe braking forces rarely occur and would cause transverse cracking or tearing. The shear stress from the tires when driving is caused by micro slipping between the tire and the AC are also small. Thermal stress can contribute to the initiation of a crack in the top surface. But it is the value of ΔK, which is not dependent on the daily thermal variation that causes fatigue as pointed out previously. To produce a tensile stress in the AC that can cause longitudinal cracking in 3-10 years, the tensile stress would have to be at least 110 psi to neutralize the radial and tangential stresses of the order of 80 psi plus an additional tensile stress of about 30 psi to cause TDC fatigue. This appears to be unlikely.

When the truck tires are wide and the tire pressure is high coupled with a low stiffness of the AC surface compared to the stiffness of the base course, the bending stress at the bottom of the AC is small in comparison to the shear stress just inside the edge of the tire. The shear stress then becomes the dominant fatigue parameter instead of the bending stress.

The shear stress decays rapidly in the transverse direction, while the shear stress in the longitudinal direction remain high since the wheel runs along or parallel to the crack. The distribution of the shear stress is not as parabolic as it is in beam bending where the highest shear stress occurs at the mid-depth. For a typical highway, with a low modular ratio E_1^* / E_2^* (surface to base) the highest shear stress occurs at approximately the top one-quarter depth, so that there is a bias toward the top. Comparison of the shear stress at the inside edge of the tire and bending stress at the base of the AC surface, directly beneath the wheel load is presented in Figure 7.13. It shows that the shear stress is much greater than the bending stress. Thus the TDC problem offers intellectual stimulation and becomes one of mounting concern. What is the mode of cracking and what is the rate of crack propagation? It must clearly be associated with the relatively large shear stress, so that ***the mode of cracking is the tearing MODE* III** of fracture mechanics.

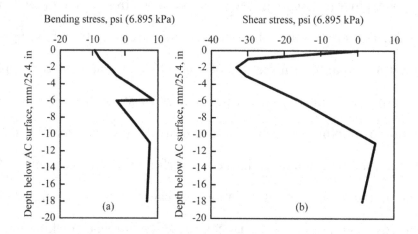

Figure 7.13. Comparison of the maximum shear stress and bending stress in the AC

The stress intensity factor for Mode I am obtained by analogy with Mode III as

$$K_{III} = \tau \frac{\sqrt{c\pi}}{1 + 0.125\lambda^2 c^2} \qquad (7.20)$$

where τ = the maximum shear stress.

In the Dugdale model the crack-tip opening displacement in Mode I is $\delta_T \approx K_I^2 / EY$, where Y = initial yield stress. By analogy the crack tip opening displacement in the tearing Mode III is

$$\delta_T = \frac{K_{III}^2}{EY} \qquad (7.21)$$

and the rate of crack propagation becomes

$$\frac{dc}{dN} = 0.153 \frac{K_{III}^2}{EY\left(1-K_n^2\right)} \qquad (7.22)$$

where $K_n^2 = K_{III}^2 / K_{IC}^2$.

The crack initiates at the top because the shear stress is higher in the upper layer as shown in Figure 7.13. The initiation of the crack is aided by the thermal curl stress and a lower shear strength caused by segregation (Roessler, 2007). The ratio of the shear stress and the bending stress does not increase with increase in the tire pressure. It decreases from 9.7 to 4.3 in the Example Problem 7.3 presented below. Therefore the increase in the tire pressure for the same load does not appear to contribute to TDC.

7.12 Stabilized materials

Several materials can be treated with lime, cement or bitumen to improve their strength and durability. Lime is especially useful for clayey soils by reducing its plasticity and expansibility, and increasing its workability. Cement is used for almost any type of soil to increase strength and durability. Bituminous stabilization may be used hot or cold for improving the strength by adhesion, waterproofing and decreasing the swelling potential. Bitumen does not coat the particles of aggregate, but form globules that adhere to the aggregates. The type of soil used in bituminous stabilization of materials is presented in Table 7.4.

For pavement design, the parameters that are needed are the tensile strength, modulus of elasticity, and fracture toughness. Each material must be tested for its properties including aging, anti-stripping, resistance to ultraviolet light, the amount of VOE (volatile oxidation emissions) before acceptance for design. Aging of asphalt is caused by oxidation at the molecular level. It can be beneficial by producing a stiffer material or it can be harmful if aged so much that the material becomes too brittle.

The challenge in physical property characterization is to develop tests that describe key asphalt binder parameters and how they change throughout the life of the asphalt concrete. At present, the best way is to determine the optimum level of aging by testing its tensile strength, fracture toughness and complex modulus and then evaluating quantitatively the fatigue life and rutting by computer programs, such as *CONPAVE, ALLIGATOR* or *RUT.*

Table 7.4 Materials suitable for stabilization with bitumen.

% Passing Sieve (in.)	Sand-Bitumen	Soil-Bitumen	Sand-Gravel
1-1.5			100
1	100	100	60-100
No. 4	50-100	50-100	35-100
No. 10	40-100	35-100	13-50
No 100			8-35
No. 200	5-12	Good-3-20	
		Fair—0-3, &20-30	
		Poor >30	
Liquid Limit		Good-<20	
		Fair—20-30	
		Poor 30-40	
Plasticity index		Good-<5	
		Fair—5-9	

7.13 Cement-treated and asphalt-treated materials

There is a scarcity of data giving the correlation between the tensile strength and the modulus for stabilized materials. Some of the data are presented graphically below.

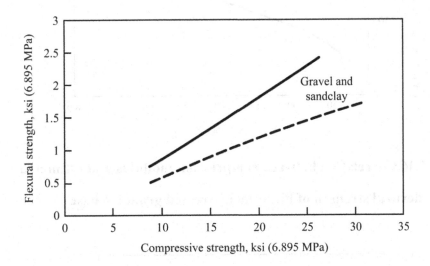

Figure 7.14. Flexural strength as a function of compressive strength of cement-treated base (Croney, D. & P., 1989)

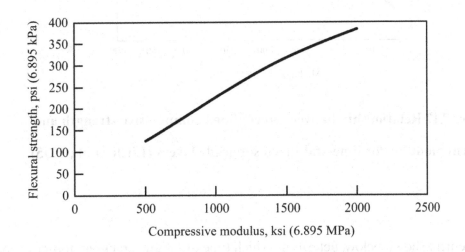

Figure 7.15 Flexural strength as a function of compressive resilient modulus, cement treated base

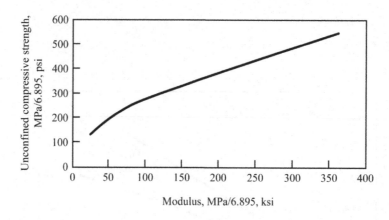

Figure 7.16.Correlation between compression modulus and estimated

flexural strength of bituminous-treated granular base

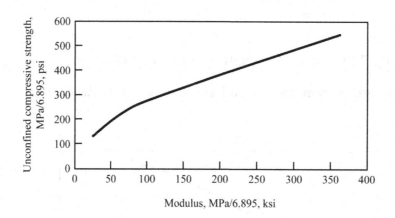

Figure 7.17 Relationship between unconfined compressive strength and

resilient modulus for lime-stabilized subgrade layers (Little et al., 1995)

Example 7.3

For the pavement shown below, determine which type of fatigue cracking, top-down or

bottom-up dominates and the number of cycles to failure. Tests show that the upper AC

has some segregation causing 20% lower shear strength. The tensile strength of AC is

405 psi and the shear strength is 350 psi.

40 kN (552 kPa)
↓↓↓↓↓↓↓↓↓

AC surface	$E_1^* = 689$ MPa	$\nu_1 = 0.35$	$H_1 = 15.2$ cm
Bit. Stab.	$E_2 = 402$ MPa	$\nu_2 = 0.35$	$H_2 = 30.5$ cm
Clay Gravel	$E_3 = 276$ MPa	$\nu_3 = 0.40$	$H_3 = 15.2$ cm
Sandclay	$E_4 = 62$ MPa	$\nu_4 = 0.40$	

Figure 7.18 Cross section of Pavement with Bituminous Stabilized Base

Solution

Determine the max thermal stress caused by a temperature drop of $10^o C$. The transverse axial force per unit width for a uniform temperature drop of $10^o C$ is

$$N = E^* \frac{\alpha}{1-v} h\Delta T 88.91 \; kPa \; (12.9 \; psi)$$

where $E^* = 689.5 \, MPa \, (100 \; ksi), h = 15.24 \; cm, \alpha = 5.5(10^{-5})/^o C, \Delta T = 10^o C$

The thermal bending moment per unit width is

$$M_T = E^* \frac{\alpha}{1-v} \frac{h^2}{6} \frac{\Delta T}{2}$$

Therefore the additional bending stress for a temperature differential from top to bottom of -10 deg. C is

$$f_b = 6Mh^2 = E^* \frac{\alpha}{1-v} \frac{\Delta T}{2} = 487 \; kPa \; = \; (70.7 \; psi)$$

The combined thermal stress is 12.9 + 70.7 = 93.3 psi (643 kPa). The thermal stresses have a frequency of only 1 cycle per day, so that thermal fatigue is small relative to that caused by the shear stresses when considering fatigue. In fracture analysis it is considered as the minimum stress in the computation of the value of ΔK for cyclic loading. The thermal stress

105

also adds to the bending stress in causing fracture, which occurs mostly at low temperatures and can contribute to *the initiation of the crack* in the top surface.

Table 7. *CHEVRON* stresses for pavement shown in Fig. 7.18

Layer	Modulus. psi	Tensile strength.,psi	Shear strength,.psi
AC surface	100,000*	405	350
Bit. stab. base	70,000*	380	510
Clay gravel	96,000	205	230
Sandclay subgrade	9000		10

Unit conversion: 1psi = 6.895 kPa

Effect of the tire pressure on the shear stress and the bending stress

To study this effect, the tire pressure was varied for the same pavement from 80 to 120 psi (827 kPa), keeping the wheel load at 9000 lb. (552 kN). The corresponding shear and bending stresses were found from *CHEVRON.* The results are presented graphically below.

Evidently increasing the tire pressure does not contribute to the occurrence of TDC. However, the effect of the increasing the stiffness of the tire walls should cause more TDC.

In the pavement shown above, the shear stress $\tau = 25.1\,\text{psi}$ is much higher than the bending stress $f_b = 2.58\,\text{psi}$, *because of the low ratio of the stiffness of the AC layer and that of the base, which is the primary cause of top-down fatigue cracking.* The rate of crack propagation is expressed as

$$\frac{dc}{dN} = 0.153\frac{1}{EY\left(1-K_n^2\right)}\left(\Delta K_{III} - \Delta K_{th}\right)^2$$

where $K_n = \left(K_{III} / K_{IIIc}\right)^2$, and

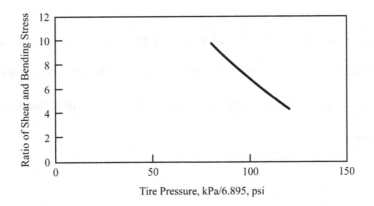

Figure 7.19 Effect of the tire pressure on the shear stress/bending stress, $E_1^* / E_2^*, = 1.42$

$$K_{III} = \frac{\tau\sqrt{c\pi}}{1 + 0.125\lambda^2 c^2}$$

where $K_{IIIc}^2 = \dfrac{1}{1+v} K_{Ic}^2$.

The data from Fig. 7.7 are input to **ALLIGATOR** which gives the number of cycles for

fatigue cracking to grow from the top down as shown below

Table 7.5 Top-down crack growth with the number of cycles of load

Location of crack	No. of 80kN ESALS millions	Crack length, in (cm) Transverse	Longitudinal
Primary	14.4	24.6 (62.4)	27.3 (69.3)
Secondary	19.7	20.48 (52.0)	32.8 (83.3)

The length of the crack may be about 18 inches shorter than that at the surface. The

longitudinal crack spacing depends of the statistical spacing of the inherent defects. The

longitudinal cracks have the tendency to coalesce and form one long crack because they are

in the wheel paths, whereas each transverse crack is crossed only once per vehicular passage.

Example 7.4

Determine the fatigue life of the pavement shown below. The base course may be constructed of **either** *15 cm of soil-cement of 15 cm of bituminous stabilized sandy gravel*. The material properties of the base are as follows: The same sandy gravel is used in either case.

40 kN (552 kPa)

↓↓↓↓↓↓↓↓↓↓

AC surface	$E_1^* = 1379$ MPa	$H_1 = 15.2$ cm
Cement treated or Bit. Stab. base	$E_2 = $??? MPa	$H_2 = 20$ cm
Sandclay	$E_3 = 103$ MPa	

Figure 7.20 Cross-section of flexible pavement with a stabilized base

Cement-Treated sandy-gravel-base

Modulus (bending) $E_2^* = 4758$ MPa

Tensile strength $f_b = 364$ psi $(2509$ MPa$)$

Fracture toughnes $K_{Ic} = 569 \, psi\sqrt{in}. \left(kPa\sqrt{m} \right),$

Fatigue Limit $f_e = 60(10^{-6})E_2^*$

Bituminous stabilized sandy-gravel base

Modulus (bending) $E_2^* == 855$ MPa$, v_1 = 0.35$

Tensile strength $f_b = 2234$ kPa

Fracture toughness $K_{Ic} = 539$ kPa psi$\sqrt{in}.$

Fatigue Limit $f_e = 60(10^{-6})E_2^*$

Subgrade-silty-sand

Modulus $E_3^* = 103425$ MPa$, v_3 = 0.40$

Solution

Cement treated base

CHEVRON stress analysis

Maximum bending stress at bottom of base course	$f_b = 352\,kPa$
Modulus of subgrade reaction	$k = 92567\ kPa\,/\,mm^3$
Inverse characteristic length	$\beta = 0.015\,cm^{-1}$
Fracture toughness	$620\ kPa\sqrt{m}$

The data shown above are put into **CONPAVE** together with the stresses from which gives the following output:

No of ESALS at 500,000 daily:

Primary transverse crack at approx. 3.66 m	5.06 million cycles
Secondary and corner cracks	21.5 million cycles

The material length $l_0 = \dfrac{K_{Ic}^2}{f_b^2} = 7.2$

Therefore the fatigue life of the pavement with cement treated base course is about 21 million cycles with 500,000 ESALS per year Allowing for reliability of 78% and 95 %confidence level, the fatigue life is estimated to be approximately 16 million cycles, or 32 years.

Bituminous –treated sandy gravel

Maximum bending stress at bottom of base course	$f_b = 238\,kPa$
Endurance stress at 60 μ	$f_e = 103\,kPa$
Modulus of subgrade reaction	**k = 553 pci**
Inverse characteristic length	$\beta = 0.034\,cm^{-1}$.
Fracture toughness	$486\,kPa\sqrt{m}$)

No of ESALS at 500,000 daily:

Primary long. and transverse cracks	1.8 million cycles
Secondary and alligator cracks	4.28 million cycles

The characteristic length $l_0 = \dfrac{K_{lc}^2}{f_b^2} = 4.16$, so that fatigue life and the crack spacing are both too small. Before revising, check rut depth of the base course.

Rut depth—Bituminous Stabilized Sandy Gravel

The following material properties are the input to the computer program'

From the ε_v^p *vs.* $\ln p$ *curve*, $\lambda = 110$

Triaxial shear strength test, steady state: $c_f, \phi_f = 345$ kPa, 40 deg.

The stresses and elastic vertical strain obtained from **CHEVRON** are:

Top 10 cm of base at midpoint (z = 20 cm)

$\sigma_1 212$ and $\sigma_3 = -20\ kPa$

$\varepsilon_z = 0.000162$

Bottom 10 cm of base at midpoint (z = 30.5 cm

$\sigma_1 = 69$ kPa and $\sigma_3 = -143$ kPa

$\varepsilon_z = 0.000091$

Total rut depth for base = 1 cm. This value is somewhat too large, since the total rut depth of the pavement should be smaller than 2.54 cm. The bituminous mixture should be modified by adding more bitumen to increase the cohesion and tensile strength. The analysis should then be repeated!

7.14 Advantages of the analytical fracture mechanics analysis and design

1. The theory provides a rapid means of quantitative evaluation of new materials, such as **EVAPAVE** for pavement design. It permits optimization of the design process for engineering efficiency and business economy.

2. The rational design methodology is formulated in terms of the *material properties, the geometry and boundary conditions.* It accounts for the varying thermal stresses and temperature-dependent properties of materials that exist in the field.

3. There is no scaling factor. The performance of any multilayered system, flexible or rigid can be predicted with sufficient accuracy for practical purposes.

4. It is simple, easy to use and imparts to the designer a feel for the design process.

5. It facilitates the accurate tracking of structural distress in a pavement, and condition assessment for maintenance.

6. Recycled EVA is less expensive than AC or PCC and helps to clean up the environment. Most of all, it will have **no joints, no cracks no bumps nor depressions**, assuming that the subgrade is strong with negligible rutting. Accordingly, it will save fuel and vehicular maintenance costs and will be the smoothest riding pavement ever.

7. With a rational structural design, only a few parameters are required to be determined experimentally.

8. With a rational structural design, there is little need for experimental test pavements. It is possible to go directly from laboratory testing of the material properties to construction of the highway.

9. It captures the fundamental behavior of the structural design of pavements with sufficient accuracy for practical purposes.

PROBLEMS

7.1 The material properties and maximum bending stress in the unnotched beam are:

Compressive strength (28 day)	34.47 MPa
Simply supported span	3.0 m
Depth	45.0 cm
Width	22.5 cm
Depth of notch at midspan	1.25 cm
Maximum bending stress	46.0 kPa

Determine (a) the fatigue life for haversine pulses at 10 cps, pulse time of 0.1 s.

(b) What should be the minimum dimensions of the beam to satisfy ASTM requirements for a valid fracture toughness determination? Use the computer program **CONBEAM** for fatigue.

7.2 The material properties and maximum bending stress in an asphalt concrete beam are:

Indirect tensile strength	350 kPa
Dynamic complex modulus in compression	2200 kPa
Simply supported span	1.0 m
Depth	20 cm
Width	10 cm
Depth of notch at midspan	0.65 cm
Maximum bending stress	90 kPa

Determine (a) the fatigue life for haversine pulses at 10 cps, pulse time of 0.1 s.

(b) What should be the minimum dimensions of the beam to satisfy ASTM requirements for a valid fracture toughness determination? Use the computer program **ACBEAM** for fatigue.

7.2

The material properties and maximum bending stress on EVAPAVE indirect tensile strength specimen and a beam are:

Indirect tensile strength	5000 kPa
Dynamic complex modulus in compression	$14(10^6)$ kPa
Simply supported span	3.0 m
Depth	1.5 m
Width	1.5 cm
Depth of notch at midspan	1.5 cm
Maximum bending stress	360 kPa

Determine (a) the fatigue life for haversine pulses at 10 cps, pulse time of 0.1 s.

(b) What should be the minimum dimensions of the beam to satisfy ASTM

(c) What should be the minimum dimensions of the beam to satisfy ASTM requirements for a valid fracture toughness determination? Use the computer program **ALLIGATOR** for fatigue.

7.3

The cross section of a flexible pavement is:

AC viscoelastic surface modulus = 800 MPa; elastic modulus = 2400 MPa; thickness = 15 cm

Bit. stab. base modulus = 500 MPa; thickness = 25.4 cm

Silty gravel subbase modulus = 250 MPa; thickness = 15 cm

Silty subgrade modulus = 72 MPa

Fracture toughness of the AC = 680 kPa $\sqrt{m}$, yield strength = 1800 kPa

Calculate (a) the fatigue life of the AC surface

8

RUTTING IN FLEXIBLE PAVEMENT SUBGRADES

Rutting of flexible pavement subgrades has received a fair amount of attention. A mechanistic model was developed and verified experimentally in the laboratory and in the field for subgrade soils by Majidzadeh (1976, 1978). Several empirical methods of estimating the amount of rutting by been proposed by Allen and Deen (1986), Saraf *et al* (1986), Townsend and Chisholm (1976) and Dorman and Metcalf (1965). The last reference is the one most frequently used. It relates the number of cycles of loading to cause rutting failure as a function of the initial elastic deflection, using multilayer elastic theory. The main shortcoming of this method is that the amount of rutting is strongly dependent on the subgrade drainage and pore-water pressures generated and dissipated that are not considered.

8.1 Analytical model for rutting

An analytical model was developed by Ramsamooj and Alwash (1990) and Ramsamooj et al. (1999). The model utilized multiyield surfaces with isotropic and kinematic hardening. It describes the behavior of saturated soils under cyclic loading in terms of the material properties, the loading and boundary conditions. The model was validated using laboratory cyclic tests for sand. It was extended to handle stress-induced anisotropy and pore-water pressure generation and dissipation (Ramsamooj and Piper, 1992). It can predict the rutting of a flexible pavement subgrade under repeated vehicular loading for various conditions of drainage, undrained to fully drained.

The yielding of soils which exhibit anisotropy that is symmetrical about the vertical z-axis using cylindrical coordinate system (r,θ,z) may be described by elliptical surfaces of the form given by (Sandler et al. 1976: Prevost 1978):

$$f = (\sigma_z - \sigma_r - 1.5\alpha_z^m)^2 + c^2 \left(p - \beta^m\right)^2 - \left(k^m\right)^2 = 0 \qquad (8.1)$$

where $\alpha_r^m = \alpha_\theta^m = -\alpha_z^m / 2$ and $1.5\alpha_z^m$ and β^m are the centers of the yield surfaces in the deviatoric plane and the hydrostatic axis, respectively, σ_r and σ_θ are the effective normal stresses, p = the mean effective normal stress, k^m = the instantaneous size of the mth yield surface and c is initially assumed to be $\left(3/\sqrt{2}\right)^{1/2}$. The yield surfaces define regions of constant elastoplastic moduli as shown in Figure 8.1(a) and (b) for the deviatoric plane (Prevost, 1978; Scott 1985). Only elastic changes occur inside the yield surface, whereas plastic deformations occur for all stress paths directed toward the exterior of the yield surfaces.

The general plasticity relations are as follows. Let $g(\sigma)$ be the plastic potential. If $\mathbf{Q}$ and $\mathbf{P}$ are the vectors normal to the yield and potential surfaces, respectively, then &&& (8.2) where the vectors may be resolved into deviatoric and dilatational components, respectively:

$$\mathbf{Q} = \frac{\partial f}{\partial \sigma} : \qquad \mathbf{P} = \frac{\partial g}{\partial \sigma} \qquad (8.3)$$

The non-associated flowrule gives the plastic strain

$$d\varepsilon^p = L \frac{\partial g}{\partial \sigma} \qquad (8.4)$$

For axisymmetric loading, the value of P'' is determined as follows. Consider the response of a soil in an undrained triaxial test when subjected to a deviatoric stress increment $\Delta\sigma_1 = \sigma_1 - \sigma_3$, in the vertical direction. A pore-water pressure Δu is generated. The volumetric strain consists of two parts, namely the elastic or recoverable part and the plastic or permanent part. From Eqs. (8.4) and (8.5).

$$d\varepsilon_v^p = \frac{d\sigma_1}{H'k}P'' \qquad\qquad (8.5)$$

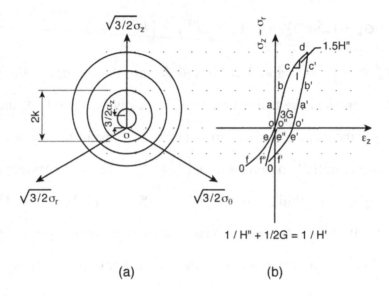

$$\frac{1}{H''} + \frac{1}{2G} = \frac{1}{H'}$$

(a)   (b)

Fig 8.1 Triaxial tests (a) stress space representation and (b) stress/strain curve

For deviatoric loading, the elastic and plastic strains in the vertical direction are respectively

$$d\varepsilon_1^e = \frac{d\sigma_1}{3G} \qquad\qquad d\varepsilon_1^p = \frac{2d\sigma_1}{3H'} \qquad\qquad (8.6)$$

where G = the elastic shear modulus and H' = the plastic modulus. The vertical strain in the subgrade is

$$d\varepsilon_z = \sum_{(m=1)}^{n}\left(\frac{1}{3G} + \frac{2}{3H'^{(m)}}\right)(d\sigma_1 - d\sigma_r)^m \qquad\qquad (8.7)$$

where σ_z and σ_r = the vertical and radial stresses and m is the *mth* yield surface. Cyclic loading may induce anisotropy because of structural or fabric changes in the soil, which depend on the density and on the extent to which the particles are elongated or platy.

8.2 The hardening and softening rules

Arthur and Menzies (1972) found that changes in the shear strength and modulus of sands caused by structurally-induced anisotropy are about the same when the results are normalized in terms of effective stress ratios and maximum strain. Changes in stress orientation relative to fabric orientation influence volume change tendencies This, in turn, influences the dilatancy contribution to the strength of the sands and the volume changes in drained deformation and the pore-water pressures in undrained shear strength of clays. Arthur et al. (1977) found that anisotropy in sands could be induced by a rotation of the principal stress directions in repeated preshearing. Ishibashi et al. (1988) found that when the stress is sufficiently high, the fabric induced by the previous stress maximum in the opposite direction is totally erased. Furthermore the new fabric induced by shearing in this direction would be nearly the same as if the sample had been sheared monotonically from an anisotropic condition to the stress level without any previous stress history.

In a pavement the direction of the principal stress axes rotates as the vehicular load passes over a point. However, the stresses decay rapidly as the wheel load moves away from the point, so that the vertical and horizontal stresses directly under the point remain the dominant major and minor principal stresses. Accordingly, the changes in the model parameters are made at the end of each cycle of loading. During the loading cycle α_z is constant for both drained and undrained loading, whereas β moves with the stress point, so that $\beta = p$ for all yield surfaces (Prevost, 1978), where p is the effective mean pressure. During undrained loading, two changes take place, namely there is a plastic volumetric strain while simultaneously there is an increase in the pore water pressure reducing the effective stress and decreasing the volumetric strain, so that the total volume remains constant.

The outermost yield surface is a volumetric yield surface. It expands and contracts with increase or decrease in the void ratio, in accordance with the critical state model concepts of Roscoe and Burland (1968). For each yield surface expands or contracts the same amount.

The increment in the size of each yield surface is (Prevost 1978, Ramsamooj and Alwash 1990) is derived from

$$\frac{dk^p}{k^p} = \frac{dp^p}{p^p} = \lambda \varepsilon_v^p \qquad (8.8)$$

Therefore

$$dk^p = dk^m = k^p \lambda \varepsilon_v^p \qquad (8.9)$$

Similarly $\qquad\qquad dp^p = dp^m = p \lambda \varepsilon_v^p \qquad (8.10)$

where k^p, p^p = the initial size of the outermost yield surface along the deviatoric and hydrostatic yield surfaces, respectively, ε_v^p = the plastic volumetric strain caused by cyclic loading, and $\lambda = \lambda_c - \kappa$ is the slope of the permanent volumetric strain ε_v^p versus $\ln k^p$ ($\lambda = $ d($\ln p$)/d ε_v^p) in an undrained triaxial test as shown in Figure 8.2.

To account for the hardening either the shear moduli or the size of the yield surface can be varied (Ramsamooj and Alwash 1990). For each type of triaxial test, drained, undrained with pore-water pressure dissipation, or undrained, the elastic wall (Fig.8.2) is the same. However, the pore water pressure generated is different. This means that although the amount of hardening is the same, the amount of softening is not; for example in the drained test the pore water pressure is zero, so there is no softening as in an undrained test.

Ramsamooj and Alwash (1990) showed that in an undrained triaxial test, the pore water parameter associated with the mth yield surface does not vary during cyclic loading from its value in the first cycle. This relationship holds providing there is no redistribution of the water content or structural changes in the soil

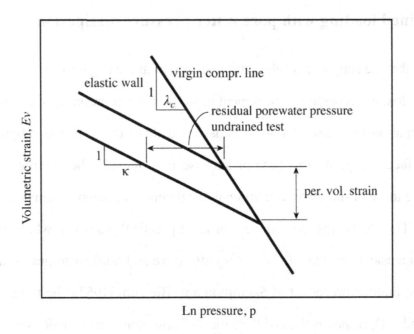

Figure 8.2 Idealized volumetric stress/ strain curve

There is also a small volumetric strain caused by a single load or by cyclic loading due to change in the effective mean pressure. It is given by:

$$d\varepsilon_{vp}^{p} = \frac{3(1-2v)}{K} dp \qquad (8.11)$$

where $K = E/(3(1—2v))$ = bulk modulus. For constant amplitude loading in which σ_1 and σ_3 are constant, changes in the mean effective pressure are caused by changes in the pore water pressure. The permanent volumetric strain in the first cycle can be measured by a triaxial shear test or a hydrostatic pressure test, loading and unloading. Alternatively, they may be obtained from cyclic tests using 100-1000 repetitions, depending on the magnitude of the applied loads.

The preceding analysis enables the prediction of the stress path or stress-strain response of sands under cyclic loading, since changes in the stress path and the stress-strain response can be fully accounted for by changes in the sizes of the yield surfaces and the kinematic rule for undrained loading.

8.3 Undrained loading with pore water pressure dissipation

It is assumed that the subgrade is fully saturated such as at the beginning of the spring season. The vehicular loading generates an increment in the pore water pressure, while simultaneously there is dissipation by consolidation. It is assumed that the top of the subgrade is a free drainage surface if a permeable base or subbase is provided. The loading and boundary conditions are those of the three-dimensional (3-D) consolidation problem solved by Gibson *et al.* (1967). The solution is too complex for most practical purposes, whereas the solution to the one-dimensional problem (1-D) is very attractive and familiar to practicing engineers. Following the famous precedent of Skempton and Bjerrum (1957), the time factors *T* and *T'* for 1-D and 3-D, respectively, and the ratio *T/T'* are plotted as functions of the degree of consolidation in Figure 8.3 (Gibson et al. 1967). The graph shows that the ratio is constant and equal to 3.70, for *a/H* =1.0, where a = the radius of the loaded area and *H* is the thickness of the compressible stratum. This means that the calendar time *t* in the time factor **T** for the solution for the 1-D problem can be replaced by 3.70*t* in the time factor for the 3-D problem.

The solution for the pore water pressure in the classical 1-D consolidation is given in most textbooks on geotechnical engineering. A vehicle driving at a rate of 50 km/hr. generates an excess pore water pressure under undrained conditions, corresponding to a pulse loading of

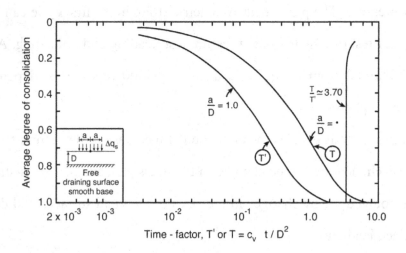

Figure 8.3 Consolidation of stratum under circular load

about 0.1 second. As the wheel passes over a point in the pavement the elastic portion of the pore water pressure reduces to zero immediately after the passage of the wheel load. Consequently the excess pore water pressure is zero at the depth of the subgrade, below which the deviator stress is elastic. The pore water pressure distribution may be assumed to be parabolic or triangular with the top of the subgrade being the highest, because the deviator stress is highest. Either a parabolic or a triangular distribution of pore water pressure may be assumed. However, in the illustrative problem that follows, it is shown that the pore water pressure generated in the subgrade during undrained loading is approximately triangular.

For a triangular distribution of pore water pressure with drainage at the top, the boundary condition at the bottom is $u_i = 0$, when the depth $z = H_{dr}$ the length of the longest drainage path or the distance from the top of the subgrade to the depth where the deviator stress is elastic. The excess pore water pressure is

$$u = \sum_{n=1}^{n=\infty} A_n \sin\frac{n\pi z}{H_{dr}} e^{-n^2\pi^2 T} \tag{8.12}$$

where the time factor $T = 3.70 c_v t / H_{dr}^2$, where c_v = the coefficient of consolidation. For a triangular distribution of pore water pressure $u_i = u_t(1 - z/H_{dr})$ and the boundary condition gives $u = u_i$ when $T = 0$. Therefore the pore water pressure at a depth z at any time t will decrease from its initial value due to consolidation according to

$$u\left(\frac{z}{H_{dr}}\right) = \frac{2}{H_{dr}} \int_0^{H_{dr}} \left(u_i\left(1 - \frac{z}{H_{dr}}\right)\sin\frac{n\pi z}{H_{dr}}\right) dz \sin\frac{n\pi z}{H_{dr}} e^{-n^2\pi^2 T} \tag{8.13}$$

Therefore

$$u\left(\frac{z}{H_{dr}}, T\right) = \sum_{n=1}^{\infty} \frac{2u_t}{n\pi} \sin\frac{n\pi z}{H_{dr}} e^{-n^2\pi^2 T} \tag{8.14}$$

The internally generated pore water pressure u_g for a large number of cycles of a relatively small deviator stress under undrained conditions increases approximately linearly with time (Seed 1976, Stoll and Kald 1976), so that

$$u_g = at = \psi T \qquad (8.15)$$

where a is a constant, $\psi = H_{dr}^2$ multiplied by the rate of generation of the excess pore water pressure under undrained conditions, determined experimentally.

The linear increase in u_g can be combined with the dissipation by the superposition integral (Christian 1976) giving a residual pore water pressure of

$$u_r\left(\frac{z}{H_{dr}}, T\right) = u_g(0)u\left(\frac{z}{H_{dr}}, T\right) + \int_0^T u\left(\frac{z}{H_{dr}}, T - \tau\right)\frac{du_g}{dt}d\tau \qquad (8.16)$$

The ultimate buildup of the pore water pressure $(t \to \infty)$ at the midpoint of the subgrade undergoing permanent deformation is

$$u_r\left(\frac{1}{2}, \infty\right) = \frac{\psi}{8} \qquad (8.17)$$

In other words, the buildup of the pore water pressure at the midpoint of the subgrade, allowing for both generation and dissipation of the pore water pressure by consolidation, cannot exceed one–eighth of the rate of generation in an undrained triaxial test. Assuming that this is a desirable limit for the pore water pressure means that $\psi/8 = p$, where $p =$ the confining pressure shows that the value of the coefficient of consolidation

$$c_v = \frac{H_{dr}^2}{8(3.70)t_g} \qquad (8.18)$$

where $t_g =$ the time for generation of the pore water pressure in an undrained triaxial test subjected to and simulating the stress pulse in the subgrade generated by the passage of a vehicular wheel load travelling over the pavement.

8.4 Changes in the position of the water table

The model can handle changes in the moisture of the subgrade because of changes in the position of the water table, providing the water table is within 6 m from the top of the subgrade. This depth is the maximum depth at which the principles of soil suction have been found to be applicable by Croney (1958). If the water table is at a depth of h cm below the top of the subgrade, the suction is

$$s = \log_{10}\left(\sigma + \gamma_w h\right) \tag{8.19}$$

where σ = the effective stress (kg/cm^2) and   is the density of water (kg/cm^2). Therefore the change in the soil suction due to a rise or fall in the water table can be easily be obtained. The corresponding change in the water content w of the subgrade at this depth can be obtained from the *experimental* suction-moisture content curve (Figure 1.4).

Having determined the total changes in the pore water pressure from vehicular generation and dissipation (Eq.8.18) and the change in the water content from rise or fall of the water table, it remains to determine the total change in the water content or the equilibrium moisture content of the saturated soil. This can be determined from the relationship given by Bear (1973) as:

$$\partial w = \frac{1+v}{G_s}\partial\varepsilon_v \tag{8.20}$$

where ε_v is the total volumetric strain, u = the pore water, G_s = the specific gravity of the soil solids, k = the permeability, v = the void ratio of the soil and t = calendar time.

What is the effect of the change in the water content on the permanent volumetric strain shear strength of the soil? This is vital to the prediction of the permanent deformation or rutting and the probability of shear failure in the subgrade and the pavement as a whole. The permanent volumetric strain is obtained from Rowe's stress dilatancy theory which states that for triaxial compression, the ratio of the permanent volumetric strain and the permanent vertical strain is

$$\frac{d\varepsilon_v^p}{d\varepsilon_1^p} = \left(1 - \frac{\sigma_1}{K(\sigma_3)}\right) \tag{8.21}$$

where $d\varepsilon_1^p$ = the permanent axial strain, determined experimentally, σ_1 and σ_3 are the effective major and minor principal stresses, and $K = \tan^2\left(45^o + \frac{\phi_f}{2}\right)$. This relationship will be elaborated on in Chapter 9. The effects of water content and pore water pressure on the performance of soils are a well-known in soil mechanics. No aspect of soil mechanics deserves more consideration. Terzaghi once said "water is the culprit" in most failures of dams, highways and other foundations. Simply stated the shear strength at failure is given by

$$\left(\sigma_1 - \sigma_3\right)_f = \left(\sigma_1 + \sigma_3\right)\sin\phi_f + 2c\cot\phi_f \tag{8.22}$$

But it is far from simple!

8.5 Changes in the subgrade stresses with repeated loading

Repeated loading causes permanent vertical and volumetric strains in the subgrade. To predict the stress path for cyclic loading would mean that the variation of each parameter would have to be predicted cycle after cycle. Cyclic loading causes strain hardening, which means that the material properties to change. For high cycle rutting, it is simpler to change the modulus of the outermost yield surface to account for the net hardening. Ramsamooj and Piper (1992) showed that Skempton's pore pressure coefficient $A = du/d\sigma_1$ for any yield surface is constant. The model parameters are modified at the beginning of the succeeding cycle as summarized below:

Triaxial drained cyclic loading

Shear vol. strain

$$d\epsilon_v^p = d\epsilon_{v0}^p \left(1 + \lambda \epsilon_v^p\right),$$

Mean pressure vol. strain

$$d\epsilon_{vp}^p = d\epsilon_{vp0}^p / \left(1 + \lambda \epsilon_{vp}^p\right),$$

Total volumetric strain

$$d\epsilon_{vp}^p = d\epsilon_{vp}^p + d\epsilon_v^p,$$

Triaxial undrained cyclic loading

In addition to the above equations for drained cyclic loading, the Skempton A parameter and the residual pore water pressure are, respectively

$$\frac{du}{d\sigma_1} = \text{constant}$$

$$u_r\left(\frac{z}{H_{dr}}, T\right) = u_g(0) u\left(\frac{z}{H_{dr}}, T\right) + \int_0^T u\left(\frac{z}{H_{dr}}, T - \tau\right) \frac{du_g}{dt} d\tau$$

$$u_r\left(\frac{1}{2}, \infty\right) = \frac{\psi}{8}$$

8.6 Experimental validation of the model predictions

Permanent deformation in a drained cyclic triaxial test on loose Ottawa sand

The model predictions were also verified for drained cyclic loading on No. 30 Ottawa sand. The model parameters obtained drained triaxial hydrostatic tests are presented in Table 8.1, with a listing of the material properties at the bottom. The sand was subjected to 12 cycles of loading

TABLE 8.1. Drained triaxial test on loose sand

m	Vertical stress $\Delta\sigma_1$, psi	Per. vol. strain $\varepsilon_v^p (10^{-3})$	Vert. strain (10^{-3})
1	0	0	0
2	4.07	0.125	0.020
3	8.14	0.292	0.005
4	12.21	0.521	0.095
5	16.28	0.855	0.205
6	20.35	1.689	0.410
7	24.42	1.689	0.83
8	27.69	1.802	1.25

Material properties: Void ratio $e = 0.97$, $p = 20$ psi , k^p =28 psi, $\lambda = 204, \varnothing = 33^o$

with a deviatoric stress of 27.69 psi (190.9 kPa) at a rate of loading of one cycle per second.

The test data for the deviatoric stress versus the vertical strain are presented in Figure

8.4 and the corresponding prediction by **SUBGRADE** (Appendix VI) in Figure 8.5. The

agreement between the theory and the experiment is good.

Permanent deformation and pore water pressure buildup under cyclic triaxial loading on Newfield clay

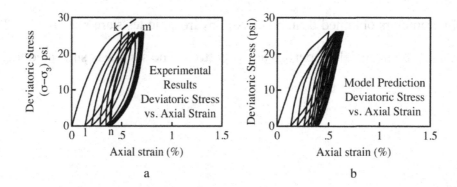

a

b

Figure 8.6 Comparison of predicted and experimental deviator stress vs. axial strain

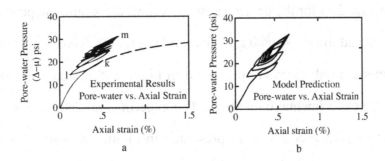

Figure 8.7 Comparison of predicted and experimental

pore water pressure vs. axial strain

The test data for the first cycle, loading and unloading is given in Table 8.2. The soil was normally consolidated to 57 psi (392 kPa). The critical stress level defined as the maximum stress level at which the soil can be cycled without failure was 35 psi. For test B1-T22 (Sangrey 1968) and test T3 (Sangrey se al., 1969) the cyclic stress level was 26.2 psi. The deformation and pore water pressure increased until a maximum was reached after 11 cycles. Additional cycles of loading produced no further net change and the stress-strain and pore water strain formed closed hysteresis loops. This was called the non-failure equilibrium state. From the test data in all of the model parameters were obtained. The void ratio e $wG_s = \kappa \ln 57 = 0.62 =$ where $= \kappa =$ the rebound compression index. The compressibility index is equal to the virgin compressive index less the rebound index $\lambda = \lambda_c - \kappa$

$$\lambda = \frac{d \ln p}{d \varepsilon_v^p} = 204 \qquad (8.23)$$

The sensitivity of the soil is not given, but judging from its liquidity index and salt concentration it is estimated to be about 6 (Mitchell 1976). The shape of the stress path of the first cycle of the experimental data **in** Figure 8.8b indicates that the soil structure changed significantly. The test data indicate that the unloading is not elastic as there is some plastic flow resulting in unclosed hysteresis loops. The model could handle soil structural changes, so that the predictions were limited to the third to the eleventh. When plastic flow occurs in unloading it is handled in the same way as for loading.

The experimental data for the deviatoric stress versus the strain, the pore water pressure vs. the axial strain, and the stress path are presented in Figures. 8.6a and b, respectively. The corresponding model predictions are presented in Figures. 8.7a and 8.7b, indicating good agreement with the experimental data above. The experimental data for the stress path and the corresponding model prediction are presented in Figures 8.8a and 8.8b, respectively, showing good agreement.

Table 8.2 Undrained triaxial test on Newfield Clay (Sangrey, 1968)*

m	Vertical stress $\Delta\sigma_1$, psi	Pore water pressure u, psi	Vert. strain ε_y, 10^{-3}
Loading			
1	0	0	0
2	4	0.125	0.020
3	8	0.292	0.005
4	12	0.521	0.095
5	16	0.855	0.205
6	20	1.689	0.410
7	24	1.689	0.83
8	26.2	1.802	1.25
Unloading			
2	22.2	23.27	5.36
3	18.2	21.67	5.06
4	14.2	20.32	4.70
5	10.2	19.07	4.16
6	6.2	18.02	3.60
7	2.2	17.32	2.56
8	0	17.03	1.96

* Second cycle (test B1-T22 or T3). Material properties: $p = 57$ psi, $k^p = 54$ psi, $\lambda = 40.3$.

Table 8.3 Experimental data from Sangrey et al. (1969), Sangrey et al. (1968)

Liquid limit	28%	Clay fraction	36%
Plastic limit	18%	Hydrous mica	40-60%
Natural water content	24%	Chlorite	20-40%
Salt conc.	8 g/L	Calcite	0-20%
Coefficient of consolidation	$0.1\ in^2$ / min.	Compression index	0.12
Specific gravity	2.75	Virgin comp. index λ_c	0.0512
Rebound comp. index K	0.0108	Angle of friction $\emptyset$	29 deg.

The *overconsolidation ratio defined as the ratio of the preconsolidation pressure* p_c' *and the current consolidation pressure is*

$$OCR = \frac{p_c'}{p_0'} = 1 + \lambda \varepsilon_v^p \qquad (8.24)$$

From the OCR the pore water pressure coefficient at failure A_f is determined from the experimental data. The model prediction for the OCR after 11 cycles of loading is 2.82. From This value of the OCR corresponds to a value of $A_f = 0.08$. Therefore the pore water pressure at failure is equal to the residual value at the end of the 11th cycle (22.58 psi or 155.7 kPa) plus 0.08 times the deviator stress at failure. Since the angle of internal friction at failure is 29^0, the deviatoric stress at failure after 11 cycles of loading is 56 psi (390 kPa) as shown in Figure 8.11. The predicted deviatoric stress at failure agrees closely with the experimental result of 57 psi (593 kPa).

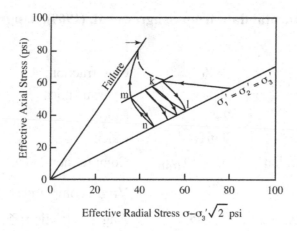

Figure 8.8a. Experimental data for effective axial stress vs. effective radial stress (Sangrey et al. 1976, and Eagan and Sangrey, 1978)

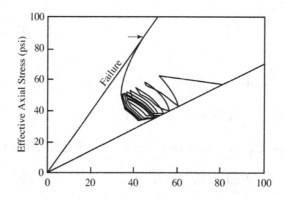

Figure 8.8b Model prediction for effective axial stress vs. effective radial stress (data from Sangrey et al. 1976, and Eagan and Sangrey, 1978)

8.7 Application to rutting of a flexible pavement subgrade

The computer program *SUBGRADE* determines the elastic and plastic vertical strains, the subgrade deviator stress and the dynamic modulus as functions of the number of applications of vehicular load, as well as the rate of generation of pore water pressure in undrained loading. It also computes the total rut depth due to elastic and plastic volumetric and shear deformation, as well as the vertical deformation due to densification and the excess pore water pressure generated. A flexible pavement consisting of 15.24 cm of asphalt concrete (AC) over a silty clay subgrade as shown in Figure 8.9 was used. The AC properties were obtained from Majidzadeh et al. (1976), and those of the silty clay subgrade were determined from triaxial and consolidation tests in the Soil Dynamics Laboratory, California State University, Fullerton. The basic material properties are presented in Table 8.4, and the undrained triaxial data on the silty clay for the stress, displacement and pore water pressure are presented in Table 8. 5. The distribution of the deviator stress with depth for a 53.4-kN wheel load is shown in Figure 8.9. At a depth of 72 cm below the top of the subgrade, the stress is smaller than the elastic limit, so that rutting will not occur below this depth.

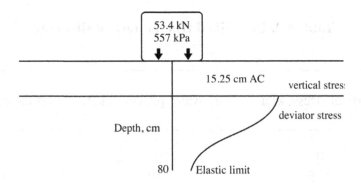

Figure 8.9. Deviator stress vs. depth below subgrade

The material properties for the asphalt concrete and subgrade are given in Table 8.5

Table 8.5. Material properties of asphalt concrete and subgrade

Comp. layer	Material property	Asphalt concrete	Subgrade
AC surface:	Young's modulus, MPa	1,236,735	62.5
	Poisson's ratio (assumed)	0.3	0.4
Subgrade:	California Bearing ratio (CBR)		6.0
	Dry density, kN/m^3		16.1
	Void ratio		0.64

SUBGRADE prediction of the permanent vertical strain versus the number of cycles of loading is presented in Figures 8.10 and 8.11 for wheel loads of 40 and 53.4-kN, respectively. The results show that the permanent deformation in the subgrade after one million passages of vehicular loading of a 40-kN and a 53.4-kN wheel load under drained conditions is less than 0.1 cm, and 0.2 cm, respectively.

For undrained conditions (rate of dissipation of pore water pressure much slower than the rate of generation of pore water pressure by vehicular loading), the pavement would fail by excessive permanent deformation after 250 cycles and 220 cycles for a 40-kN and a 53.4-kN wheel loads, respectively, if the loads were applied at the rate of 140 cycles per day. The prediction for the residual pore water pressure for a 53.4-kN wheel load, for $c_v = 180 \, \text{cm}^2 / \text{day}$

Table 8.4. Undrained triaxial test on silty clay

m	Vertical stress, kPa	Pore water pressure, kPa	Vertical displacement, cm
1	0	0	0
2	27.5	5.91	0.0056
3	42.2	9.26	0.0087
4	55.0	12.91	0.0129
5	68.7	16.21	0.0185
6	82.4	18.89	0.0269
7	96.2	21.15	0.0403
8	137.4	26.79	0.0884

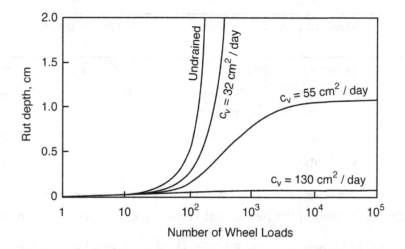

Figure 8.10 Rut depth vs. number of 40 kN wheel loads

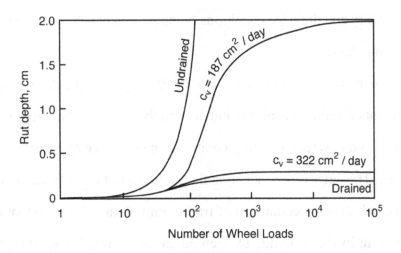

Fig 8.11 Rut depth vs. number of 53.4 kN wheel loads

and $\psi = 728$ kPa are presented in Fig. 8.12. The results confirm the theory that the buildup

of the pore water pressure, when drainage is provided at the top of the subgrade, cannot

exceed one-eighth of the rate of generation of the pore water pressure, as expected. The

ratio of the rut depth caused by densification and the total rut depth is dependent on the

drainage conditions. For undrained conditions it varies from 20% to almost zero as failure is

approached, whereas it varies from 20% to 56% for a 53.4-kN load when there is adequate

drainage.

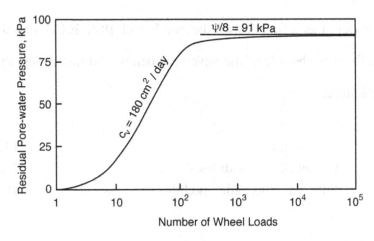

Figure 8.12. Residual pore water pressure vs. number of 53.4 kN wheel loads

The theory can account explicitly for most of the important variables of the subgrade, such

as the stress/strain/strength characteristics, the permeability and compressibility of the soil, the loading, boundary and drainage conditions of the pavement subgrade, and changes in the position of the water table.

The preceding analysis shows that the rate of development of rutting is dependent not only on the amount of drainage available but also on the ability of the subgrade soil itself to dissipate the excess pore water pressure generated in the subgrade by repeated applications of vehicular loading. The latter is dependent on the coefficient of consolidation of the subgrade soil. The theory indicates the conditions of traffic, rainfall and drainage that can cause the pore water pressure in the subgrade to accumulate to a value higher than the confining pressure causing failure. Specifically this would happen when the coefficient of consolidation of the subgrade soil $c_v < H_{dr}^2 / (29.6t_g)$, where t_g is the time for the pore water pressure to reach the confining pressure in an undrained triaxial test. Calling this the critical value C_{vc}, of it is clear that the pore water pressure will not exceed the confining pressure and that the rut depth will reach an equilibrium only if $C_v > C_{vc}$. In other words, no amount of drainage by the provision of a permeable subbase can be effective, if the coefficient of consolidation of the subgrade is inadequate. *This is an important new finding!*

Example 8.1

Determine the rut depth as a function of the number of 18-K ESALs applied at a rate of 250,000 per year, for the subgrade of the pavement below, and investigate the possibility of progressive shear failure.

Pavement	AC: modulus =	1379 MPa psi
	Gravel Base : modulus =	276 MPa
	Silty clay Subgrade: modulus =	41.3 MPa

Solution

CHEVRON stress analysis:

$\sigma_1 = 20.5\,\text{kPa}\,(\text{comp}).,\sigma_3 = -11.7\ \text{kPa},(\text{comp})\varepsilon_z = 0.000331$

Other properties:

Permeability of the AC surface $k = 9 \dfrac{cm}{day}$; $G_s = 2.72$; $e_0 = 0.64$; $w = 0.195$

H_{dr} = distance from top of subgrade to depth where the stresses are elastic 30.5 cm

$E_4^* = 41.3 \text{ MPa}$, $E_{4e} = 276 \text{ MPa} \left(\text{tangent or elastic modulus} \right)$

Rowe's stress-dilatancy parameters: $K = 2.95, c_f = 20.7\, kPa, \phi_{cv} = \phi_f = 30^o$,

Exptl. data: Hydrostatic compression: $c_v = 0.64\, cm^2 / min$

Rate of generation of pore water pressure (av. of first 1000 cycles) $a = 0.28$ kPa/min. Since the time factor $T > 1.0$ for full dissipation by consolidation, the minimum time for full dissipation of the excess pore water pressure is only 40 minutes, corresponding to 200 vehicles. After that the rate of dissipation of the excess pore water pressure (pwp) by consolidation will be equal to the rate of vehicular rate of generation of the pwp, so that the excess pwp will remain constant. This value is given by $u_{max} = \psi / 8 = 110.2 / 8 = 13.8 \text{kPa}$.

Rainfall: Assume that the rainfall in the rainy season greater than 2 in/day The infiltration into the pavement is controlled by the permeability of the AC or the rainfall intensity, whichever is smaller. Therefore the infiltration is 2 in./day. The pore water pressure generated in the subgrade at the end of this day is $u = 62.4(1/6)/144 = 0.072$ psi/day = 3.96 psi $\dfrac{\psi}{8} = 0.5\, psi$ for 55 days. Allowing for dissipation by consolidation the maximum pore water pressure is per rainy season. Assume for simplicity one jump per year.

The above information is put into the computer program **SUBGRADE** which gives the number of cycles for local shear failure to occur as $N = 500, 000$ cycles, with the maximum pore water pressure being 2.0 psi and the rut depth in the subgrade 3.3 cm. $\leftarrow Ans$.

8.8 Liquefaction of sands

Sometimes a highway may be constructed over a loose sand deposit. Such a subgrade may be susceptible to failure by liquefaction under a strong earthquake. Liquefaction is also involved in the joint pumping problem of PCC pavements. An elastoplastic model is presented for predicting the displacement and liquefaction of sands in for both monotonic and cyclic loading. The model utilizes multi-yield surfaces, isotropic hardening for drained lading and a combination of isotropic and kinematic hardening for undrained loading as described previously.

The state of the art of predicting the number of cycles of loading for liquefaction to occur has made great strides since 1958, when Casagrande first introduced his concept of the critical void ratio. If the sand has a void ratio greater than the critical void ratio, liquefaction can occur. However, Seed (1967) found experimentally that sand of any density can liquefy. Casagrande (1975) disagreed with Seed's conclusion, stating that what was observed in the laboratory compression triaxial tests on dense sands was not true liquefaction, but cyclic mobility caused by the migration of pore water to the regions of stress concentration at the top and bottom caps. This presentation supports Casagrande concepts and presents a mathematically simple model that captures the fundamental behavior of sands with sufficient accuracy.

The yield criterion of soils is represented by regions of constant elastoplastic moduli as shown in Figs. 8.1 (a) and 8.1(b) for the deviatoric plane. Only elastic changes of deformation occur for stress paths inside the yield surface, while plastic deformations occur for all stress paths directed toward the exterior of the yield surface. The model accounts for inherent anisotropy, but not stress-induced anisotropy. It uses isotropic hardening for drained loading so that the yield surfaces may expand or contract the same amount about their respective centers. For drained loading, the centers of the yield surfaces do not move. For undrained loading, a combination of isotropic and kinematic hardening is used so that the p-coordinate of all the centers of the yield surfaces traversed by the stress point moves with the stress point. The outer yield surfaces move only after the stress point has crossed the inner yield surface except if it is being pushed by the inner yield surface. The yield surfaces, therefore, do not intersect at any point (Prevost, 1978).

8.9 Model parameters

Sand is characterized by conducting a compression drained triaxial test, an extension drained triaxial extension test, and a hydrostatic compression test, loading and unloading. The stress/ strain curves and the stress space representation and are presented in Fig. 8.1(a), from which the values of α^m, H^m, G^m and are found. The yield surfaces are chosen to define regions of constant total shear moduli the superscript m refers to the region between the mth and the $(m+1)^{th}$ yield surfaces. The slope of the stress-strain curve gives the elastic and plastic moduli and the corresponding elastic and plastic strains, or

$$d\epsilon_1^m = \frac{d\sigma_1^m}{2G} + \frac{d\sigma_1^m}{2H^m} = d\epsilon_1^{em} + d\epsilon_1^{pm} \qquad (8.26)$$

in which $d\epsilon_1^{em}$ and $d\epsilon_1^{pm}$ = the elastic and plastic volumetric strains, respectively.

8.10 Liquefaction and cyclic mobility of sand in triaxial compression

Casagrande (1975) *defined liquefaction as the state of stress when the pore water pressure increases to the point where the shear strength is lowered below the applied shear stress, causing failure with unlimited deformation, and zero tendency to volume change.* It follows that for liquefaction to occur in a triaxial compression test, $d\epsilon_v^p / d\epsilon_1^p$ must be equal to zero. Therefore

$$\sigma_{1-u} = K(\sigma_3 - u) \qquad (8.27)$$

Furthermore Eq. (8.27) is also the condition for steady-state failure with the stress point on the failure envelope. Therefore *liquefaction occurs with unlimited deformation and zero tendency for volume change.*

In dense sand, on loading the pore water pressure increases initially, because there is a positive change in the volume until point C (Figure 8.13) is reached.

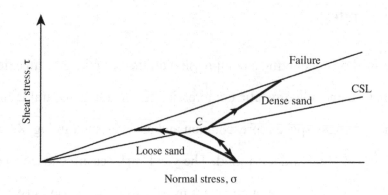

Figure 8.13 Stress paths for dense and loose sands

Here conditions for liquefaction are not satisfied. The stress path then follows an upward trend towards the failure envelope given by the friction angle $\varnothing_{cv}$. However, further loading causes a strain hardening effect that enables the sand to carry load until it collapses at a very large strain. Therefore *dense sand does not liquefy in monotonic loading,* but may develop very large strains (Castro, 1975, Seed, 1978).

8.11 Cyclic response of saturated sands

Sand strain hardens due to compressive volumetric strain and strain softens when the confining pressure is decreased due to the buildup of the pore water pressure. When dealing with an undrained compression or extension loading, the strategy adopted is as follows. A load increment $d\sigma_1$ is imagined to be applied in a triaxial test with full drainage, so that there is a volume change $\delta\varepsilon_v^p$. Simultaneously the confining pressure in the drained triaxial test is imagined to be reduced by an amount, equal to the increment in pore water pressure du, causing an equal and opposite volume change $-\delta\varepsilon_v^p$ so that the net volume change is zero.

The plastic volumetric strain in a drained triaxial test is caused mainly by shear straining. The relationship between the volumetric strain (elastic + plastic) and the confining pressure in a drained triaxial hydrostatic consolidation test with rebound was presented in Figure 8.2. The slopes of the compression and rebound lines are λ_c and κ, respectively. The residual pore water pressure at the end of the first cycle is δu_r, so that the effective confining pressure is $\sigma_3 - \delta u_r$. The combined hardening and softening is given by

$$k- \rightarrow k\left(1+ \lambda\delta\epsilon_v^p\right) \qquad (8.28)$$

in which $\lambda = \lambda_c - \kappa$, and ϵ_v^p = net plastic volumetric strain. from the first to the nth cycle, given by

$$\epsilon_{vn}^p = \sum_1^n \delta\epsilon_{vn}^p \qquad (8.29)$$

Since the rebound compression line is straight, the residual pore water pressure from the first the nth cycle is

$$u_n = \sigma_3\left(1-e^{-\kappa\epsilon_{vn}^p}\right) \qquad (8.30)$$

The pore water pressure caused by the *elastic* compression of the soil skeleton is (Ramsamooj, 1998):

$$\delta u_1 = \frac{\delta\sigma_1}{3}+1-\sigma_3\left(e^{\kappa\epsilon_{v1,}^p}\right) \qquad (8.31)$$

$$\delta u_n^m = \sum_1^m \frac{\delta\sigma_1^m}{3} +1- \sigma_3\left(\kappa\sum_1^m\epsilon_{vn}^{pm}\right) \qquad (8.32)$$

Therefore the entire stress path can be predicted for an undrained test from the drained test results

Example 8.1

For the pavement shown below, determine the rut depth and the possibility of shear failure as a function of the number of 80 kN ESALs applied at a rate of 250,000 per year.

Pavement

AC surface: $E_1^*\,(dyn.mod.in\,bending) = 1200\,\text{MPa}, H_1 = 15$ cm

Silty gravel base: $E_3^* = 360$ MPa, $H_3 = 20$ cm

Sand subbase: 360 MPa, = 20 cm

Silty clay subgrade: $E_4^* = 50\,\text{MPa}; \text{E}$

Elastic modulus =276 MPa

Subgrade soil properties

Hydrostatic comp.: Lambda $\lambda = 35$

Rebound $\kappa = 110$

Void ratio $e = 0.62$

Coeff of consol. $c_v = 2\,in^2\,/\,day\,(6.45\,cm^2\,/\,day$

Shear strenght: $c_f = 15$ kPa, $\varnothing_f = 30^o\,(\text{steady-state})$

Specific gravity $G_s = 2.72, w = 0.195$

Rowe's stress-dilatancy parameters $K = 2.95, c_f = 20.7\,kPa,\ \phi_{cv} = \phi_f$

Hydrostatic compression: $c_v = 51.6\,cm^2\,/\,day$

H_{dr} =top of subgrade where the stresses are elastic = 30.5 cm

Solution

CHEVRON stress analysis (Imperial units)

Stresses and strain at depth 6 in. below top of subgrade $z = 30$ in.

Vert. and hor. stresses $\sigma_1 = 1.55\,\text{psi}, \sigma_3 = 0.035\,\text{psi}$

Vertical elastic strain $\varepsilon_y^e = 0.000178$

Permanent vertical strain per cycle, first cycle (assumed) = 0.0011

Soil test data

Shear strength envelope $c_f = 5\,\text{psi}, \varnothing_f = 30^o$

Hydrostatic compression curve $\lambda = 58$

Coefficient of consolidation $c_v = 12\,in^2\,/\,day$

Elastic modulus, resilient modulus (mag).	$E_e = 45,000\,\text{psi}$
Resilient modulus (mag.)	$E^* = 9,000\,\text{psi}$
Void ratio	$e_0 = 0.70$

Residual pore water pressure after 1st cycle

(average of first 1000 cycles) $\quad\quad\quad\quad\quad\quad\quad u_r = 0.06\,\text{psi}$

Seasonal rise in the water table $\quad\quad\quad\quad\quad \Delta h$

$= 2\text{ft}$

The rainfall intensity is taken as 40 in/yr. From CHEVRON, the stresses at a depth of 36 in. are within the elastic limit, so that the length of the drainage path for consolidation is 6 in. double drainage and the thickness of the subgrade for rut depth calculation is 12 inches.

The above data are inserted into the computer program **SUBGRADE** (Ramsamooj and Piper, 1990), which gives the following output:

The total buildup of the residual pore water pressure is from vehicular loading and consolidation is 0.3 psi, that from rainfall is 0.67 psi and the increase in the water content is 3.6% after 15.96 million cycles of 18-K ESALs.

The rut depth is 0.42 in. after 15.97 million 18-K ESALs. The rut depth is reasonable. Less drainage, smaller coefficient of consolidation, less rainfall, smaller change in the depth of the water table and a weaker subgrade will lead to larger rut depth.

9

RUTTING IN FLEXIBLE PAVEMENT SURFACES

"Every great advance in science has issued from a new audacity of imagination"

John Dewey, The Quest for Certainty, 1929

9.1 Permanent deformation or rutting in pavements

The cornerstone of rational pavement design is the ability to predict the behavior of paving materials by analytical methods. Present paving practices rely on empirical procedures. Majidzadeh and Guirguis (1978) developed a procedure for rutting in Ohio pavements. Brown and Snaith (1974) studied the effects of stress, strain and temperature on the rutting of AC triaxial specimens, subject to dynamic loading for both the deviator and confining stresses. They found that the level of static confining stress that gave the same permanent strain as the repetitive loading was approximately equal to the mean level of the dynamic stress; the effect of frequency of the applied loads was dependent on the total time the load acted on the specimen, and that the effect of rest periods was negligible. All of the relationships proposed contain several arbitrary parameters that cannot be generated from fundamental principles.

New materials such as styrene-butadiene-styrene (SBS), EVAPAVE and other plastic composites demand more scientific and fundamental principles. "Ideally one may aim at the development of a model involving a large number of state variables and parameters to cover the responses of the pavements to all possible types of loading and loading histories.

However, a realistic assessment of the material properties, combined with the desire for computational feasibility calls for the development of a mathematically simple model that would capture the fundamental behavior of all paving materials with sufficient accuracy". Such a model is presented in the following.

The yield condition for soils and the corresponding equations for stresses and strains were discussed in the preceding section.

9.1 1 Triaxial compression

In the field of permanent deformation, the research work of Rowe (1971) is outstanding. Rowe used the mechanics of particulate materials to derive the theory of stress-dilatancy. For a triaxial compression test, Rowe's stress-dilatancy theory states that the ratio of the permanent volumetric $d\varepsilon_v^p$ and vertical strain $d\varepsilon_1^p$ is

$$\frac{d\epsilon_v^p}{d\epsilon_1^p} = 1 - \frac{\sigma_1}{K\sigma_3 + 2c_f\sqrt{K}} \tag{9.1}$$

where σ_1, σ_3 =effective major and minor principal stresses, respectively, $K = \tan^2(45 + \phi_f/2)$ and ϕ_f= the equivalent angle of friction between particles, modified to include simultaneous deviations of individual particles from the mean direction. The friction angle varies from ϕ_u , the angle of interparticle friction, to ϕ_σ , the critical state angle of friction. Rowe's experiments showed that $c_f \rightarrow 0$ and that the value of K was strongly dependent on the pore fluid. The plastic potential for compression of granular soils in plane strain is

$$g = \frac{\sigma_1^K}{\sigma_3} \tag{9.2}$$

For triaxial compression it is

$$g = \frac{\sigma_1^{1/(2K)}}{\sigma_3} \tag{9.3}$$

where σ_1 = vertical stress and σ_3 = the radial stress. The ratio of the permanent volumetric and vertical strain (Rowe, 1971) is:

$$\frac{d\epsilon_v^p}{d\epsilon_1^p} = 1 - \frac{\sigma_1}{K\sigma_3 + 2\sqrt{K}c_f} \tag{9.4}$$

9.1.2 Triaxial extension

For triaxial extension, the ratio of the incremental work in and the incremental work out is constant and equal to K. The ratio of the permanent volumetric strain $d\varepsilon_v^p$ and the lateral permanent strain $d\varepsilon_3^p$ is

$$\frac{d\varepsilon_v^p}{d\varepsilon_3^p} = 1 - \frac{K\sigma_3 + 2\sqrt{K}c_f}{\sigma_1} \tag{9.5}$$

where σ_1 = the *horizontal stress* and σ_3 = the *vertical stress*. From Eq. (9.4) it is deduced that the ratio of the vertical and the horizontal strains or the equivalent of Poisson's ratio (c = 0) is

$$\frac{d\varepsilon_3^p}{d\varepsilon_1^p} = -\frac{\sigma_1}{2K\sigma_3} \tag{9.6}$$

Repetitive loading by the mean pressure (very small) also cause some permanent deformation given by

$$d\varepsilon_{vp}^p = -\frac{d\varepsilon_{vp0}^p}{1 + \lambda\varepsilon_{vp}^p} \tag{9.7}$$

where the symbols have been defined previously. Therefore the total rutting consists of that caused by shear straining plus that caused by changes in the mean pressure.

9.2 Rutting in AC Surface

From Chevron multilayer analysis computer program (*CHEVRON*), the vertical, radial and tangential stresses σ_z, σ_r, and σ_t, together with the corresponding strains ε_z, ε_r, and ε_t are determined. The vertical strain $d\varepsilon_{y1}^p$ in the first cycle, after good seating of the load is determined experimentally, as the average of the first 150-1000 cycles of pulses corresponding to the highway or airport pavement loading. EVAPAVE is so hard that it may require more than 100000 cycles, and measurement by very sensitive strain gages and a data acquisition system. The volumetric strain for the first cycle is:

$$d\varepsilon_{v1}^{p} = Rd\varepsilon_{y1}^{p} \qquad (9.8)$$

From the strain hardening law the permanent volumetric strain after the nth cycle is

$$d\varepsilon_{vn}^{p} = \sum_{2}^{n} \frac{d\varepsilon_{v(n-1)}^{p}}{1 + \lambda\varepsilon_{v(n-1)}^{p}} \qquad (9.9)$$

so that permanent vertical strain is:

$$\varepsilon_{yn}^{p} == \frac{1}{R}d\varepsilon_{vn}^{p} \qquad (9.10)$$

The increment in the permanent vertical displacement is:

$$dw_{n}^{p} = \int_{0}^{H} d\varepsilon_{yn}^{p}dy \qquad (9.11)$$

where H is the thickness of the layer of AC.

The total rut depth at the end of the nth cycle is

$$w^{p} \sum_{n=1}^{N} dw_{n}^{p} \qquad (9.12)$$

9.4 *Computer program RUT* (App. VII)

The input data are as follows:

- Average permanent vertical strain determined by the first 100-1000 cycles from a compression triaxial test and a hydrostatic compression test.

- Value of the friction angle and cohesion at the critical state, c,

145

- Resilient modulus as a function of temperature

- Fracture toughness as a function of temperature

- Hardening parameter λ from hydrostatic compression tests of

The output data are as follows:

- The total permanent vertical, volumetric and lateral strains as functions of the number of cycles of loading.

9.5 Validation of the analytical predictions for rutting of AC using published data

The published test data by Morris et al. (1976), supplemented by replicate tests in the California State Laboratory, Fullerton, were used to make the predictions (Ramsamooj et al. 1999). The analytical prediction for the axial and lateral permanent strains in the triaxial compression test (Morris et al., 1974) is presented in Fig. 9.1. Good agreement in the test data exists for the axial strain at 32.2° C, but not for the lateral strain. One explanation may be that the calibration for the lateral strain device was too coarse, and another is the occurrence of micro-cracking in the specimen.

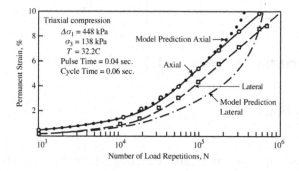

Figure 9.1. Analytical prediction for the axial and lateral permanent strains in the triaxial compression test (data from Morris et al., 1974)

146

The analytical prediction for the axial permanent strains in the triaxial compression test on Mix A, aged in the oven, (Morris et al., 1974) agrees well with the experimental data shown in Fig. 9.2. The rut depth stabilizes after about 6000 cycles, with the response tending to approach the elastic state. Brittle failure occurred in some specimens.

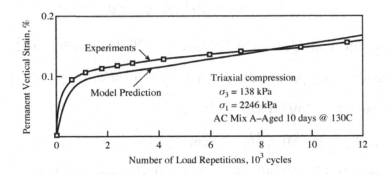

Figure 9.2. Analytical prediction for the permanent vertical strains in the triaxial compression test on aged AC (data from Morris et al., 1974)

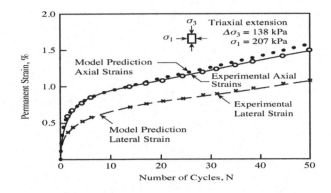

Figure 9.3. Model prediction for the permanent strains vs. the number of cycles of load (data from Morris et al., 1974)

Figure 9.3 showing good agreement. In the compression test, the ratio of the volumetric and vertical permanent strains changed from 0 to—0.15 after 330,000 cycles, the negative sign indicating volumetric expansion. In the extension test the ratio changes from 0 initially to 0.45 after 50,000 cycles.

PROBLEMS

9.1

AC surface: $E^*_{(1)}(dyn.mod.inbending) = 1250\text{MPa}, H_1 = 15\,cm$

$$c_f = 30\,kPa, \ \phi_{cv} = \phi_f = 35°$$

Hydrostatic comp.: Lambda $\lambda = 190, Rebound\ \kappa = 120,\ Void\ ratio = 0.62$

Rowe's stress-dilatancy parameters: $K = 4.0, \ c_f = 30\,kPa, \ \phi_{cv} = \phi_f = 35,$

Silty gravel base: $E^*_2 = 450\,\text{MPa}, H_2 = 15$ cm

Sand subbase: $E^*_3 = 360$ MPa, $H_3 = 15$ cm

Silty clay subgrade: $E^*_4 = 50\,\text{MPa}; E_{4e} = 276$ MPa $\left(\text{tangent or elastic modulus}\right)$

Determine the rut depth as a function of the number of 80 kN ESALs applied at a rate of 250,000 per year.

10

NEW PAVING MATERIAL EVAPAVE

There are presently 80,000 polymers on the market. A substantial amount of library research was done in developing a new material called EVAPAVE. For fatigue properties, search was made for a polymer with *high tensile strength, high fracture toughness, low melt temperature and good moisture and UV resistance.* The tensile strength was of paramount importance. The new material turned out to be a copolymer, ETHYLENE VINYL ACETATE (EVA).

10.1 ETHYLENE VINYL ACETATE (EVA)

This is a new binder invented from the preceding theories to make a new highway material called EVAPAVE. For fatigue properties, search was made for a polymer with *high tensile strength, high fracture toughness, low melt temperature and good moisture and UV resistance.* The tensile strength was of paramount importance. The new material turned out to be a copolymer, ETHYLENE VINYL ACETATE (EVA).

"EVA is a polymer that approaches elastomeric materials in softness and flexibility, yet can be processed like other thermoplastics. The material has good clarity and gloss, barrier properties, low temperature toughness stress-crack resistance hot-melt adhesive water proof properties, and resistance to UV radiation." *The free Wikipedia.*

EVA has been used in combination with asphalt cement for making asphalt concrete for many highways. However, only modest success (about 40% increases in Marshall stability) was obtained. In order to reduce the mixing temperature several tacifiers were considered, but were too expensive for practical use. However paraffin wax was low cost when combined with EVA in equal parts and effectively reduced the melting point to $185^o C$.

10.2 EVAPAVE COMPOSITION AND PREPARATION

After a few trials the proportion of each component of the materials used to produce EVAPAVE is given below.

Grading of aggregates, ¾ in. max size:	AASHO Road Test
Binder EVA: % by weight of agg.	3 and 5
Paraffin wax to reduce melting temperature by weight of agg	5%.

EVA was heated to 366 degrees F and then hand mixed with 5% paraffin wax using a large spoon. It was compacted statically to a density of 22.62 kN/m^3 (143 pcf) in a Marshall mold and extruded for testing. The resulting compacted mix is called **EVAPAVE.**

10.3 LABORATORY TESTS ON EVAPAVE

10.3.1 Tensile strength

The tensile strength with 5% EVA was obtained from indirect tension tests on 10 cm diameter x 10 cm high specimens using an INSTRON testing machine and temperatures of 5, 25 and 45° C. The tensile strength with 3% EVA was tested at 25 deg. C. The results of indirect tension tests for AC and PCC obtained from published sources are also presented in Figure 3. The bending strength and bending modulus were also obtained from simply supported beams using strain gages. The testing rate was 1000 psi/s.

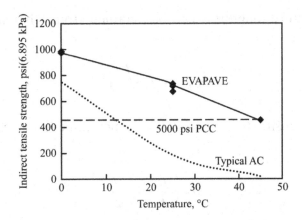

Figure 10.1. Comparison of indirect tensile strength of EVAPAVE, AC and PCC

10.3.2 Compressive strength

The compressive strength was measured by a direct compression test on the same size cylinders as stated above, using an MTS loading system at 1000 psi/s. A compressive strength at $(25^\circ C)$ of 14.3 MPa was obtained.

10.3.3 Fracture toughness

The fracture toughness test was conducted on a notched beam 7.5 cm high x 5 cm wide x

It was conducted on a notched beam 7.5 cm high x 5 cm wide x27.5 cm long. The beam notch was 50% of its height. The value of the fracture toughness so obtained was unrealistically high, and it was discarded because it did not satisfy the ASTM size requirements, which are that both the width B of the beam and the crack length at failure should c_f be 25 times the size of the plastic zone. We can now say that the required EVAPAVE beam size should be 20 cm wide x 40 cm high x 160 cm long ($L/D = 4$). Because of the high temperature and the volume of the material it would require an oven 10 times larger than our oven, a huge hot-mixer with temperature capability and a large dynamic compactor or the equivalent static loading machine. These requirements are not practicable for our laboratory or for any other university laboratory as far as we know.

However, the fracture toughness can be obtained from Kharlab's theory (**1995**) as follows:

$$K_{Ic} = f_t \sqrt{\pi \delta}$$

where f_t = the direct tensile strength and δ = the non-dimensional characteristic length of the material given by

$$\delta = \left(\frac{f_b}{f_t} - 1 \right) \frac{H}{2}$$

where H = the depth of the beam being tested.

10.3.4 Modulus of elasticity

Bending tests on beams 7.5 cm high x 5 cm wide x 27.5 cm long with very sensitive strain gages were used.

Compression modulus	13,790 MPa
Bending modulus	8.274 MPa
Bending strength	1,386 psi (9556 kPa $\sqrt{m}$)
Characteristic length δ (Kharlab)	2.87 in.

Accordingly the fracture toughness for EVAPAVE is:

$$K_{Ic} = 1386\sqrt{3.14(2.87)} = 1425 \text{ psi}\sqrt{\text{in}})(1553 \text{ kPa}\sqrt{m})$$

Example 3 EVAPAVE fatigue

Determine the fatigue life of the EVAPAVE shown in Figure 4(a) and illustrate the associated crack pattern in Figure 4(b).

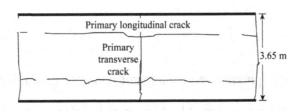

Figure 4(a) EVAPAVE cross section

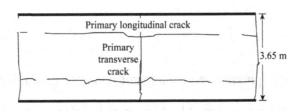

Figure 4(b) Crack pattern of EVAPAVE

Solution

The stresses are calculated with **CHEVRON**.

ALLIGATOR input data @ 25 °C

Maximum bending stress with 40 kN wheel load in wheel path	$f_b = 63\,psi\,(434\,\text{kPa})$
Indirect tensile strength	$f_{ct} = 712$ psi (4916 kPa)
Fracture toughness	$K_{Ic} = 1{,}425\ \text{psi}\sqrt{\text{in}}\ (1553\ \text{kPa}\sqrt{\text{m}}$
Bending modulus	$E_b = 8{,}274$ MPa
Endurance limit	$\sigma_{fl} = 60 E_b\,10^{-6}$

ALLIGATOR Output

Inverse characteristic length	$\lambda = \sqrt[4]{k/D} = 0.031\,\text{cm}^{-1}$
Flexural rigidity	$D = EH^3 / \left(12\left(1 - v^2\right)\right)$
Dist. of diagonal cracks from corner *along the corner angle bisector*	$= 56$ cm

153

Number of ESALS (N_1) for the primary crack to traverse the width of the pavement lane, @ 1 million cycles per year (including 43.45 million cycles for initiation of cracks)

$$N_1 = 56.5 \text{ million cycles}$$

Number of ESALS (N_2) for the secondary cracks comprising the four diagonal cracks approx. 122 cm long,, @ 1 million cycles per year

$$N_2 \gg 110 \text{ million cycles}$$

Therefore the fatigue life of the EVAPAVE pavement is practically unlimited.

10.4 Construction

The same construction equipment used for asphalt concrete is required for EVAPAVE. The asphalt plus most of the filler is replaced by EVA plus Paraffin wax. The mixing temperature of EVAPAVE is 366^oF. Cleanup after work is easier.

PROBLEMS

An EVAPAVE highway constructed of the following components:

EVAPAVE: Modulus in *dynamic compression* $E_1^* = 13800$ MPa; thickness $H_1 = 15$ cm

Low quality EVAPAVE base *dynamic compression* $E_2^* = 3200$ MPa; thickness $H_2 = 15$ cm

Sand subbase course modulus $E_3^* = 150$ MPa; $H_3 = 15$ cm

Silty sand subgrade $E_4^* = 80$ MPa

Determine the fatigue life for 80 kN axle loads applied at 1 million ESALs per day.

11

INTERACTION OF FATIGUE CRACKING AND RUTTING

Considerable complexity is involved in the interaction of cracking and rutting of a multilayer system, especially when pore-water pressures are present, necessitating the use of several areas of engineering technology. Fracture mechanics is used for fatigue crack propagation in the AC or EVAPAVE surface layer, combined with the stress-dilatancy theory and viscoelastic/plasticity for rutting. A constitutive model for elastoplastic behavior of soils, including the generation and dissipation of pore-water pressures from vehicular loading, rainfall infiltration and water table changes is used for the cyclic loading in the base and subgrade. Cracking in the surface layer results in a lower modulus thereby causing a greater amount of stresses to be transmitted to the lower layers. The increased stresses increase the amount of cracking and permanent deformation in the base and subgrade, which in turn provides less support to the surface layer, thereby generating more cracking and a continuous interaction which may eventually lead to failure of the paving system.

Fatigue cracking of the surface layer was discussed in Chapter 7. The equation for the number of cycles to cause the crack to grow from its starter value to a value of c is restated for convenience as:

$$\frac{dc}{dN} = 0.153 \frac{\left(\Delta K - \Delta K_{th}\right)^2}{E_b^* f_b \left(1 - K_n^2\right)} \tag{11.1}$$

where the symbols have been previously defined.

The rut depth is calculated separately for each layer by the following equations. The volumetric strain at the end of the n*th* cycle given by

$$\varepsilon_v^p \rightarrow \sum_1^n \frac{1}{1 + \lambda \varepsilon_{v(n-1)}^p} d\varepsilon_{v1}^p \qquad (11.2)$$

where ε_v^p = the total permanent volumetric strain at the end of the (n+1)th cycle, $d\varepsilon_{vn}^p$ = its increment in the (n + 1)th cycle, and $\lambda = d\epsilon_v^p / d(\ln p)$ curve, with p = hydrostatic compression pressure. The magnitude of the complex dynamic modulus E^* due to volumetric strain hardening at the end of the (n +1)th cycle is

$$E^* \rightarrow E^* \left(1 + \lambda \epsilon_v^p\right) \qquad (11.3)$$

The rut depth the end of the n*th* cycle is

$$H\varepsilon_{y1}^p = \sum_1^n \varepsilon_{vn}^p / \left(1 - \frac{\sigma_1}{K\sigma_3 + 2c_f \sqrt{K}}\right) \qquad (11.4)$$

where H = the thickness of the layer. When there is significant pore-water pressure in the base or subgrade, the relationship between the ratio of the permanent volumetric and permanent axial strains for triaxial conditions is

$$\frac{d\varepsilon_v^p}{d\varepsilon_y^p} = 1 - \frac{\sigma_1 - u}{K\left(\sigma_1 - u\right) + 2\sqrt{K}c_f} \qquad (11.5)$$

where u = the ambient pore-water just after construction plus the pore-water pressure generated by vehicular loading under *undrained* conditions plus that from rainfall infiltration minus that from drainage minus that dissipated by consolidation.

11.1 Vehicular rate of pore-water pressure generation and dissipation

The pore-water pressure generated by vehicular loading minus that dissipated by consolidation, or the residual pore-water pressure (Ramsamooj and Piper, 1994) is

$$u_r\left(\frac{z}{H_{dr}}\right) = \frac{4\psi}{\pi^3} \sum_{n=1}^{\infty} \sin\frac{n\pi z}{H_{dr}}\left(1 - e^{-n^2\pi^2 T}\right) \qquad (11.6)$$

where z = the vertical coordinate H_{dr} = length of the longest drainage path, $\psi = H_{dr}^2 / c_v$ = rate of generation of the pore-water pressure by vehicular loading under undrained conditions ,and $T = 3.70 c_v T / H_{dr}^2$ is the time factor for three-dimensional consolidation.

11.2 Rainfall infiltration into pavements

Rainfall infiltration through pavements or through cracks and joints in the pavements may cause softening of the subgrade and rapid deterioration of the pavement depending on the drainage provided by the underlying layers. Liu and Lytton (1984) presented a comprehensive treatment of the rainfall infiltration, drainage and load-carrying capacity of pavements. Ridgeway (1976) found that the infiltration rate into the joints of the concrete pavements in Connecticut was about 28 cc/hr./cm (0.72 cu. ft. /day/ft. of crack length). If the alligator fatigue cracks are assumed to have a spacing of 30 cm. then the equivalent permeability of the cracked pavement is 117 cm/day. When there is no ponding of water on the pavement, the seepage of water into the pavement is controlled by the intensity of rainfall or the permeability of the pavement, whichever is smaller. When there is ponding of water, the infiltration is dependent on the permeability of the pavement. It is reasonable to assume that the effective permeability is dependent on the size of the cracked area. The area of interest is that under the wheel within the radius of influence R. Accordingly the permeability of the cracked area (ft./day) is taken as

$$k_c = 0.72c = 22\,\text{cm/day} \qquad (11.7)$$

where c is the *total length* of the cracks in feet, (longitudinal plus transverse) within the radius of influence of the wheel load. The duration of ponding of the water on the pavement is assumed to be proportional to the rut depth, equal to 2 days per cm of rut depth. The seepage velocity through the pavement layer is

$$v = \frac{dh}{dt} = ki \qquad (11.8)$$

where i = the hydraulic gradient and t = time. The increment of pore-water pressure generated by the seepage in time Δt is

$$\Delta u_p = \gamma_w k \qquad (11.9)$$

When flooding occurs because the rate of infiltration is greater than the rate of drainage, the suction head at a point P in the subgrade changes from $-h_w$ to $+h_t$, =, where the former is the depth from the point in the subgrade and the latter is the height of the top of the AC above the point. Therefore the net increase in the pore water pressure is

$$\Delta u_s = \gamma_w \left(h_w + h_t \right) H_{dr}{}^2 \qquad (11.10)$$

There is also a hydrodynamic pore water pressure generated by the moving tire equal to the tire pressure at the contact surface. This pore water pressure is often assumed to be transmitted undiminished through the soil below. However, intuition dictates that this assumption cannot be realistic and that there is some lateral distribution with the depth. Unfortunately the quantitative distribution is not known presently. It will be assumed conservatively that the distribution is similar to that of the well-known Boussinesq theory. The increment in the pore water pressure is

$$\Delta u_{vp} = bt \qquad (11.11)$$

where $T = c_{vt} / H_{dr}^2$, the dimensionless time factor.

The total pore water pressure generated by the effects of rainfall infiltration or by ponding, whichever is greater, plus that from the net decrease in soil suction, plus that generated by the vehicular loading under hydrodynamic conditions, if applicable, is then

$$\Delta u = \Delta u_p + \Delta u_{ss} + \Delta u_{vp} = \psi' T \qquad (11.12)$$

Treating Δu the net infiltration is as a step function with dissipation by consolidation results in a change in the pore-water pressure given by the superposition integral as:

$$u_r\left(\frac{z}{H_{dr}}\right) = \frac{4\psi'}{\pi^3} \sum_{n=1}^{\infty} \frac{1}{n^3} \sin\frac{n\pi z}{H_{dr}}\left(1 - e^{-n^2\pi^2 T}\right) \qquad (11.13)$$

11.3 Interaction of fatigue cracking and rutting

When a crack develops in the surface layer of a pavement, its effective modulus E_{eff}^* becomes smaller offering less protection to the underlying layers. The stresses in all layers change and the viscoelastic deflection of the pavement becomes greater. The increase in the viscoelastic deflection is given by:

$$\delta w_{cycle} = \frac{2\Delta K^2\left(1-v^2\right)H}{PE^*}\delta c_{cycle} \qquad (11.14)$$

in which P = wheel load and δc_{cycle} = the increment in the crack length. The increment in the permanent displacement or rut is

$$dw_p = d\varepsilon_{y1}^p \frac{2\Delta K^2\left(1-v^2\right)H}{PE^*}dc \qquad (11.15)$$

where $d\varepsilon_{y1}^p$ = viscoplastic displacement at the end of the first cycle. The total rut depth over an interval of N cycles of loading is obtained by numerical integration by the computer program *INTERACTION* (Appendix IX) in which E^* is held constant. At the end of each interval the complex modulus is obtained by taking it as inversely proportional to w_c. The stresses in each paving layer are recalculated using **CHEVRON**. In particular the stresses in the base course and subgrade increase,11 causing more fatigue cracking and rutting of the surface. This process is iterative and was fully presented by Ramsamooj (1994).

11.1 Example of Fatigue and Rutting Interaction

A scenic coastal highway with pavement design shown below **in Figure 11.2** is expected to carry 3 million 80 kN ESALS at a rate of one million per year. Analyze this pavement for fatigue and rutting and for interaction of fatigue and rutting

40 kN (552 kPa)
↓↓↓↓↓↓↓↓↓↓

AC $E_1 = 3102$ MPa $\nu_1 = 0.15$ $H_1 = 30.5$ cm

$K_{Ic} = 623$ kPa $\sqrt{m}$

Sandy clay subgrade $E_2 = 103.4$ MPa $\nu_2 = 0.3$

Figure 11.2. Coastal Highway Pavement Design

Fatigue AC properties

Bending modulus = 0.42(3102) = 1302 MPa

Initial pore water pressure after first cycle (triaxial test) = 3.45 kPa

CHEVRON computer output

Bending stress = 400 kPa

σ_1, σ_3 = major and minor prin. Stresses = 27.8 and 2.2, resp.

Vert. strain at z= 46 cm = 0.000253

Modulus of subgrade reaction = 16.4 N/cm^3

ALLIGATOR computer output

Fatigue Cracking

No. of ESALS for *primary crack* to propagate 76 cm and 178 the transverse and longitudinal directions, respectively. Rate of loading = 250,000 80kN ESALs per year.

Rut depth after 1.66 million 80 kN ESALs   =3.45 cm.
Pore water pressure buildup after 1.66 million cycles   =8.5 kPa

INTERACTION Computer Output

Fatigue Cracking and Rutting

No. of ESALS for *primary crack* to propagate 30 in and 70
in the transverse and longitudinal directions, respectively   = 1.36 million

Rut depth after 1.36 million 18-K ESALS   =4.17 cm

Pore water pressure buildup after 1.36 million cycles   =27.1

No shear failure occurs, but continued loading beyond the formation of both primary and secondary cracks would probably cause shear failure in a weaker subgrade.

PROBLEMS

Determine the rut depth in the AC layer for the pavement below. You may make reasonable assumptions that must be completely stated.

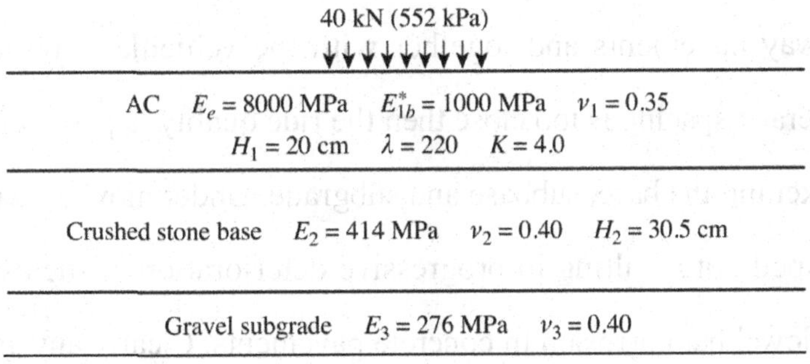

40 kN (552 kPa)

↓↓↓↓↓↓↓↓↓↓

AC $E_e = 8000$ MPa $E_{1b}^* = 1000$ MPa $v_1 = 0.35$

$H_1 = 20$ cm $\lambda = 220$ $K = 4.0$

Crushed stone base $E_2 = 414$ MPa $v_2 = 0.40$ $H_2 = 30.5$ cm

Gravel subgrade $E_3 = 276$ MPa $v_3 = 0.40$

Fig 11.1 Cross section of the pavement

12

CRACK SPACING AND THERMAL STRESS IN PAVEMENTS

Thermal stresses in colder regions such as Northern United States or Canada can initiate cracking in highway pavements and together with the vehicular stresses determine the service life. If the crack spacing is too close then the ride quality is poor, and water infiltrates the pavement weakening the base, subbase and subgrade. Under moving loads water and fine materials are pumped out resulting in progressive deterioration of the asphalt concrete or joint faulting and dowel bar corrosion in concrete pavements. Clearly any attempt to increase the spacing of the cracks is a worthwhile enterprise; an undertaking that is best achieved after the factors contributing to the spacing of cracks are understood. There have been some attempts to determine the crack spacing in asphalt concrete and concrete, one by USDOT, a theoretical approach by Li and Bazant (1995), and another by Hong et al. (1997).

12.1 Analytical solution to crack spacing in highway pavements

Hong et al. (1997) published a simple analytical model, based on fracture mechanics. They solved an important problem that considerably facilitates the design of new pavements against the thermal cracking mode of distress and the development of new materials. A brief summary of their paper follows. The pavement is modeled as an elastic plate on a Winkler foundation (Figure 12.1). However, as the main interest is transverse cracking a longitudinal strip of the plate of unit width with crack spacing of $2L$ and the crack depth $204c$ is analyzed.

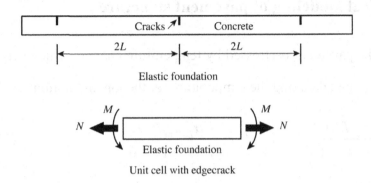

Figure 12.1. Longitudinal strip of the concrete pavement

The strip is loaded by a moments M and axial forces N at the ends. The deformation and slope of the *uncracked strip is* Δ_0 *slope* θ_0, respectively, and the deformation and slope caused by a *crack is* Δ *and* θ so that, so that the total deformation and slope caused by M and N are, respectively:

$$\Delta_t = \Delta_0 + \Delta \tag{12.1}$$

$$\theta_t = \theta_0 + \theta \tag{12.2}$$

The deformations caused by the crack may be expressed as:

$$\Delta = \lambda_{MM} M + {}_{MN} N \tag{12.3}$$

$$\theta = \lambda_{NM} M + \lambda_{NN} N \tag{12.4}$$

where $\lambda_{MM}, \lambda_{MN}, \lambda_{MN}$ = compliance functions for additional rotation, additional elongation, and elongation or rotation due to a crack of length c caused by a unit value of M, N and M or N respectively ($\lambda_{NM} = \lambda_{MN}$, for a linear plate). The SIFs are:

$$K_N = \frac{N}{\sqrt{H}} k_N \left(\frac{c}{H} \right); K_M = \frac{M}{H\sqrt{H}} k_M \left(\frac{c}{H} \right) \tag{12.5}$$

where H = the thickness of the plate, k_N and k_M, are the SIF influence values are tabulated by Tada *et al.* (1985). The functions $\lambda_{MM}, \lambda_{MN}, \lambda_{MN}$ were obtained from Tada *et al.,* but λ_{MN} required additional numerical integration by Hong et al. (1997).

12.2 Mechanical modeling of pavement structure

Assuming that the pavement is stressed by temperature change, which vary linearly across the strip thickness, and denoting the temperatures at the top and bottom as T_t and T_b,

$$M_T = \frac{E\alpha_T}{1-\upsilon}\frac{T_{t-}T_b}{2}\frac{H^2}{6} \qquad N_T = \frac{E\alpha_T}{1-\upsilon}\frac{T_{t+}T_b}{2}\frac{H}{6} \qquad (12.6)$$

in which M_T and N_T and = thermal moment and tensile force and α_T = the coefficient of thermal expansion of the concrete. The rotation due to the combined bending moment and thermal bending for the uncracked beam is

$$\theta_0 = C_{MM}\left(M + M_T\right) \qquad (12.7)$$

where $C_{MM} = 2L/D$ = compliance for rotation of the end, taken as that in a simple beam. From symmetry, the total rotation at the center is zero. Therefore

$$\lambda_{MM}M + \lambda_{MN}N + \frac{2L}{D}(M + M_T)0 \qquad (12.8)$$

If the pavement is restrained from axial contraction, then the axial force is unknown and must be obtained from the compatibility condition. The elongation of the uncracked strip is

$$\Delta_0 = C_{NN}(N + N_T) \qquad (12.9)$$

where $C_{NN} = 2(1-v^2)L/EH$ A set of coupled equations follow:

$$(C_{MN} + \lambda_{MN})M + \lambda_{MN}N + C_{MN} = 0 \qquad (12.10)$$

$$\lambda_{MM}M + \theta_0 = C_{MM}\left(M + M_T\right) \qquad (12.11)$$

If the pavement is assumed to be bonded to the subgrade and the horizontal subgrade modulus is, then the equilibrium equation in the axial direction is

$$\frac{Eh}{1-\upsilon^2}\frac{d^2u}{dx^2} - k_hu = 0 \qquad (12.12)$$

From the solution to Eq. (12.12), the compliance function for axial force at the end of the beam is

$$C_{MN} = \frac{1 - \upsilon^2}{Eh\mu} 2\, tanh\, \mu L \tag{12.13}$$

where $\mu^2 = (1 - v^2)k_h / EH$ and $\mu = 2\pi / l$, with l = wavelength and the horizontal displacement force are, respectively:

$$u(x) = C \sinh \mu x \tag{12.14}$$

$$N(x) = \frac{CE\mu H}{1 - v^2} \cosh \mu x \tag{12.15}$$

where C = arbitrary constant. Assuming that the displacement is one-half that with no restraint and $\mu = 0.05 / cm\ (0.125 / in.)$, the value of C can be found.

12.3 Crack initiation theory

The total stress intensity factor caused by M and N are

$$K = K_M + K_N \frac{M}{H\sqrt{H}} k_M + \frac{N}{\sqrt{H}} k_N \tag{12.16}$$

where $k_M (c/H)$ and $k_N (c/H)$ are obtained from Tada et al (1985). The crack initiation theory (Li and Bazant, 1994) consists of the following three postulates:

1. The maximum tensile stress σ_{max} before the initial cracks form must be equal to the tensile strength f_t of the pavement in the pre-initiation state.

2. The energy release rate G of the pavement after the initial cracks form must be equal to the fracture energy G_c of the pavement material in the post initiation state, and

3. The total strain energy must be equal to the energy cG_c to create the new crack surfaces. Bazant (1996) stated that the new fracture energy for the formation of the initial cracks G_c^{in} is smaller than the fracture energy for crack growth, or

$$G_c^{in} = \beta G_c \tag{12.17}$$

where $\beta = 0.5 \rightarrow 0.67$ (assumed to be 0.33.)

12.4 Numerical solution

Introducing dimensionless nominal stresses

$$\sigma_N = \frac{N}{f_t h}; \sigma_M = \frac{M}{f_t h} \qquad (12.18)$$

The first condition of for crack initiation is

$$\sigma_M + \sigma_N = 1 \qquad (12.19)$$

The second condition becomes

$$\sigma_N k_N + \frac{1}{6}\sigma_M k_M = \sqrt{\frac{\beta l_0}{H}} \qquad (12.20)$$

in which $l_0 = K_{Ic}^2 / f_t^2 = $ *defined as the material length of the pavement.*

The third condition becomes

$$\frac{1}{\alpha}\int \left\{ \sigma_M \, k_N(\alpha) + \frac{\sigma_M}{6} \, k_M(\alpha) \right\}^2 d\alpha = \frac{l_0}{H} \qquad (12.21)$$

Since the compliance function due to a crack is zero if the crack length is zero, the nominal stresses in the pre-initiation state

$$\sigma_M^T + \sigma_N^T = -1 \qquad (12.22)$$

Combining Eqs.(12.20) and (12.21) yields:

$$\frac{1}{\alpha}\int_0^\infty \left\{ \sigma_M k_N(\alpha) + \frac{\sigma_M}{6} k_M(\alpha) \right\}^2 d\alpha = \frac{l_0}{H} \qquad (12.23)$$

12.5 Model predictions for crack spacing in concrete pavements

The material properties of the concrete assumed by Hong et al.(1995) are:

Thickness	H= 25.4 cm (10 in.)
Modulus of elasticity	$E = 28.97$ GPa $(4.2\,(10^6)$psi;
Foundation modulus	$k_v = 108.7$ Pa/m (400 pci)

From the solution of Eq. (12.23) the material length l_0 = 17.78 cm (7 in.). Hong et al. (1995) investigated is that of pure bending without axial constraints. When the concrete is bonded to the foundation so that the axial displacement is restrained, the crack spacing becomes larger for the same values of as l_0 shown in Fig. 12.2.

From the preceding analysis, *the most important factor controlling the crack spacing in a pavement is the material length l_0.*

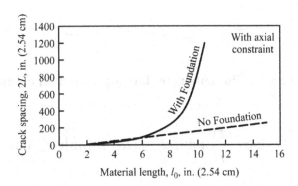

Figure 12.2 Crack spacing as a function of material length (from Hong et al (1997)

12.6 Thermal curl stresses in highway and airport pavements

Some undoweled pavements exhibit cracks at the corners of the slabs that are approximately quarter-circle in shape (Darter et al. 1955). This form of cracking results from the combined stresses caused by thermal curling, vehicular loading and moisture changes during construction. The thermal curl stress may be obtained as follows.

Consider a continuous flat slab supported by on columns as shown in Figure 12.3. If the dimensions of the slab are large in comparison **to the columns**, the deflection w is given by

$$w_p = \frac{qb^4}{384 D}\left(1 - 4\left(\frac{y^2}{b}\right)\right)^2 + \frac{qa^3 b}{2\pi^3 D}\left(A_1 A_2 - A_0\right) \qquad (12.24)$$

where $A_1 = \dfrac{(-1)^{m/2} \cos\dfrac{m\pi x}{a}}{m^3 \sinh \alpha_m \tanh \alpha_m}$

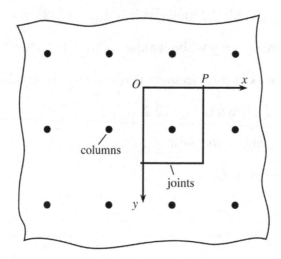

Figure 12.3 Flat concrete slab supported on columns

$$A_2 = \frac{m\pi y}{a} \sinh\frac{m\pi y}{a} \tanh\alpha_m - (\alpha_m + \tanh\alpha_m)\cosh\frac{m\pi y}{a}$$

$$A_0 = \frac{1}{m^3}\left(\alpha_m - \frac{\alpha_m + \tanh\alpha_m}{tanh^2\alpha_m}\right)$$

where $\alpha_m = \frac{m\pi b}{2a}$, a and b are the spans in the x—and y directions, respectively and $q =$ unit dead load.

If frictionless joints are placed midway between the columns, then a cantilevered slab with center support is obtained. The stresses caused by the dead load are then determined by superposing those from the continuous flat slab on those generated by the joints. The latter are determined by the **EFM** method discussed earlier.

Let there be in addition a linear thermal gradient θ, with the top of the slab colder than the bottom, so that the slab curls upward under the weight of the slab. Consider that the column is now replaced by a subgrade support with modulus equal to k, the size of the supported area to be determined later. The middle of the slab, where the weight of the slab is supported, will sink a small amount w_0. The displacement above the center of the slab at any point (x, y) due to the *thermal curl alone* is (Timoshenko and Goodier, 1951):

$$w_T = \frac{\alpha\theta}{2}\left(x^2 + y^2\right) \tag{12.25}$$

where α = the coefficient of thermal expansion of the PCC and θ = the thermal gradient. The frictional constraint is small and is neglected, conservatively. The total deflection is then

$$w = w_p + w_T + w_j + w_0 \tag{12.26}$$

where w_p = the deflection of the continuous plate caused by the weight of the slab, w_T = the deflection caused by the thermal curl, w_j = the deflection caused solely by the presence of the joints, determined by the **EFM** method, and w_0 = the deflection from the subgrade reaction caused by embedment of the slab .

The accuracy of the **EFM** method of solution was checked by comparing the deflection with that published by Timoshenko and Goodier for a 4.47 m square supported by the four corners under its own weight. Joints were placed in the continuous slab along the column lines, and the deflection obtained by **EFM** was only 2.1% smaller.

The radius R of the ground contact area is found from the condition that the weight of the slab is

$$W = 4\int_0^{b/2}\int_0^{a/2} kwx\,dx\,dy, \quad w > 0 \tag{12.27}$$

The solution of the above equations gives the values ground contact area, and W_0. The maximum bending moment for the partially supported slab with a unit weight of q is then

$$M_{max} = q\frac{a^2b}{4} - 2\int_0^{b/2}\int_0^{a/2} kwx\,dx\,dy, \quad w > 0 \tag{12.28}$$

with the maximum bending stress $6M_{max}/bH^2$

12.7 Comparison of *EFM* solutions with published results

A comparison is made of the temperature curl stresses for 202.5 cm, 254 cm, and 356 cm thick slabs with those obtained by several investigators (Table 12.1). There is good agreement between *EFM* and finite element method solutions, except for small values of the slab thickness and the subgrade modulus.

Table 12.1 Comparison of Curling Stresses with Finite Element Solutions

Slab thickness, mm	Computer model	Subgrade modulus MN/m^3		
		13.6	54.3	135.7
202.5	EFM	600	1,120	1,288
	KENSLAB*	830	1,140	1,256
	JSLAB**	648	878	920
254	EFM	706	1,242	1,418
	KENSLAB	740	1,132	1,321
	JSLAB	84	923	1,321
356	EFM	521	945	1,116
	KENSLAB	525	918	1,128
	JSLAB	414	790	977

* Darter (1977)

12.8 Bending stress caused by vehicular load with partial contact with the ground

The deflected profile of a curled concrete slab is shown in Figure 12.6. Let the deflection at the corner edge of the curled slab be w_e, and the radius of contact with the ground be R_c. . For corner loading, let the portion of the vehicular load in full contact with the ground be P_s.

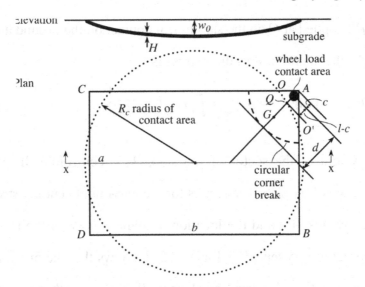

Figure 12.6 Deflected profile and plan of concrete slab on elastic foundation

The **deflection** and bending stress in the slab is treated as that caused by the load acting on the slab cantilevering from section OO', plus that caused by the load transferred to section OO', plus the deflection of A due to the deflection and slope at Q. The cantilever deflection under load P_s is

$$w_e = \frac{6P_s c\left\{l - c\left(1 - ln\frac{l}{c}\right)\right\}}{EH^3} \tag{12.29}$$

where distances c and l are as shown. The deflection w_Q and β_Q under the transferred load at Q and the transferred moment are obtained from **EFM** for a slab with free edges OC, OO, and O'B. The edge OO' is loaded with a bending moment $P(l-c)$. The total deflection at the edge is then

$$w_A = w_0 + \beta_Q l + w_e \tag{12.30}$$

The total deflection to seat the slab on the ground is taken as the load needed to cause a deflection w_A, so that $w_A - w_e = 0$, where w_e = the deflection of the corner of the slab due to the thermal curl and weight of the slab. Finally the deflection caused by the remainder of the wheel load = P- P_s , applied at Q on the slab in full contact on the ground are determined by *EFM*. Accordingly, the maximum bending stress is

$$\sigma_{max} = \frac{6}{H^2}\left\{P(d-c)+W_s(d-\rho_G)\frac{\pi}{2}k\int_0^d \rho w(d-\rho)dp\right\} \qquad (12.31)$$

where W_s = the weight of the sector of the slab, ρ_G = centroidal distance and ρ = the distance from the edge of the slab. The results for the maximum bending stresses foe a 202.5 mm slab for three values of k and the location (distance/$\sqrt{a\lambda}$) of the maximum bending stress from the corner are presented in Table 12.2. It shows that the maximum value of the bending stress caused by the vehicular load alone is slightly larger than the combined stresses of the vehicular load and thermal curl. The distance of the maximum bending stress from the corner is 2.57 $\sqrt{a\lambda}$, which is 8% longer than that given by Westergaard. In California, where the concrete pavements are not doweled, there are approximately as many pavements cracked at the midslab longitudinal edge as those that are cracked at the corner. This tends to support the solutions giving the larger stresses. More controlled experimental data are needed to assess the accuracy of the various solutions. The maximum bending stress due to thermal curl as determined by *EFM* agrees with the values from three finite element programs. The maximum bending stress for vehicular loading at the corner computed by *EFM* is about 30% larger than that given by Westergaard (1948). The distance of the maximum bending stress for vehicular loading at the corner is slightly longer than that from Westergaard.

Table 12.2 Maximum bending stresses with wheel load at the

corner, with and without thermal curl, 202.5 mm slab

Loading		Subgrade Modulus, MN/m^3		
124-kN axle load		13.6	54.3	135.7
Veh. Load Only				
EFM max. Bending stress		2,530	2,296	2,124
Distance(λ/a)		2.60	2.65	2.47
WES* max. Bending stress	1,848	1,606	1,413	
Distance (λ/a)		238	2.38	2.38
Veh. *Load + Temp. Curl*				
EFM max. Bending stress		2,427	2.075	1,848
Distance (λ/a)		2.41	2.44	2.23

- Westergaard(1948)

The maximum bending stress for vehicular loading at the corner, combined with the thermal curl at the longitudinal stress for a 202.5 mm thick slab, is slightly smaller than that for vehicular loading alone, which is approximately equal to that at the longitudinal edge. The downward thermal curl at the longitudinal edge causes an increase in the bending stress at the edge, so that the critical location for initiation of fatigue cracking is at the middle of the longitudinal edge (with the wheel load tangential to the edge) and not at the corner.

Example 12.1

Determine the maximum bending stress in a 202.5 mm thick concrete pavement, 6.09 m long and 3.66 m wide due to a thermal gradient of $33°C/m$, and the deflection caused by thermal curl, and vehicular loading as a function of the radial distance from the center. The concrete has a 90-day compressive strength of 27.5 MPa (4000 psi) with a modulus of elasticity of 27.6 GPa, and a subgrade modulus of 31.0 MPa. The coefficient of thermal expansion is 2.9 $(10^{-6})/°C$.

Solution

- Subgrade coefficient k = 13.6 MN/m^3

- Deflection due to thermal stress (Eq. 9)) = $9.52\,r^2$

- Dead load deflection of the continuous slab (Eqs. (0) and () = $1.05\left(r/R\right)^{0.66}$ mm

- Deflection caused solely by the joints (**EFM**) = $0.54(rl/R^2)$ mm

- Deflection from both wheel loads (**CHEVRON**) = $0.0472 - 0.0075\left(R - r\right)^{0.36}$

- Radius of the contact area = 3.29 m

- Maximum bending stress at the center of the slab = 600 kPa ← Ans

13

REFLECTIVE CRACKING IN RIGID PAVEMENT OVERLAYS

The structural requirements of an asphalt concrete overlay on a rigid pavement are severe. The overlay must be continuous over the Portland cement concrete (PCC), which has a joint every 500 cm. Furthermore, there are large stresses in the overlay from vehicular loading and thermal changes that cause unbonding of the overlay from the PCC. Many different types of overlays have been used and evaluated by trial and error methods (Frederick, 1984, Button 1989, but only limited success has been achieved; the overlays crack after 1-6 years irrespective of the thickness used. The products tried include full-width nonwoven and woven fabrics, rubberized asphalt treatments, reinforced asphalt strips, composite fiberglass-asphalt binder systems, stress – relieving interlayers (SRI), stress-absorbing membrane interlayers (SAMI), etc.

Frederick (1984) reported results of SRI trials on highways made at five locations in 1980-1981 in the state of New York. He stated that "the results of annual surveys have yet to prove conclusively the value of using these materials or their competitive effectiveness". In Texas, several types of geotextile fabrics have been installed at four locations. According to Button, 1989, "results based solely on these test pavements, indicate geotextile fabrics are not cost-effective in addressing reflective cracking." However, these studies used thin fabrics and not the thickened fiberglass composites over the joints that are proposed herein.

In modern times there have been unprecedented advances in materials such as thermoset and thermoplastic composites that offer the designer a wide choice of design properties. For an overlay, a high tensile strength and a low modulus is required. The E-glass/polyester composite has a relatively high tensile strength/modulus ratio and it is the least expensive. The heart of any design is in the stress analysis. At present, the author is unaware of any analytical solutions for the stresses in an overlay over a jointed rigid pavement. Due to the complexity caused by the joints, only numerical solutions have been attempted so far. The finite element method is powerful, but requires considerable skill in dealing with joint problems, and the results are often inaccurate and must be viewed with care (Huang, 1993). Fracture mechanics is an effective tool, since joints can be analyzed as "designed cracks". An analytical method of determining the stresses and crack propagation of reflective cracks in terms of the loading, boundary and environmental conditions is presented. The stresses and deflection of the composite overlay are obtained from fracture mechanics, using the relationship between the deflection and the SIF for a crack or joint. All modes of cracking are analyzed. The thermal, bending and shear stresses together with the number of cycles for repeated loading for the first reflective crack to occur are presented.

13.1 Design of the AC/Fiberglass (ACF) overlay on a rigid airport pavement

The cross section of an airport is presented in Figure 13.1. The PCC is 45.7 cm thick with joint spacing of 7.62 m and the ACF overlay is 2.54 cm thick. Over the joints the E-glass polyester mat (fiberglass, FG) is 2.54 cm thick and is gradually tapered from a length of 20 cm from the joint to a thickness of 0.19 cm within a distance of 30 cm. For adequate skid resistance the FG is manufactured with a skid proof surface. To ensure a flat level surface, an AC overlay is constructed over the FG with the paving done in the transverse direction, using FG sheets that tapered to fit the geometry of the ACF as shown in Figure 13.2, which gives the design. The material properties of the AC and ACF overlay are compared in Table 13.1.

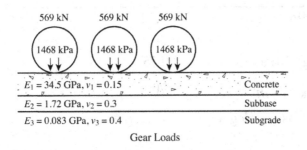

Figure 13.1 Design of the airport pavement cross section

The design of the ACF overlay is shown in Figure 13.2. The thickness of the ACF varies from 2.5 cm to 0.19 cm.

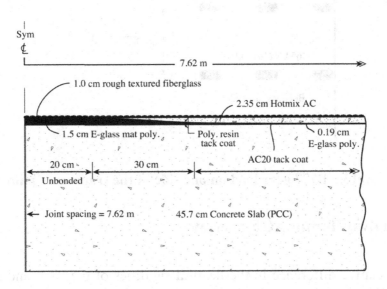

Figure 13.2 Design of the fiberglass overlay

Table 13.1 Material properties of AC and ACF overlays

Material	Asphalt concrete	Fiberglass (FG)
Complex modulus, MPa	3.44	17,238
Poisson's ratio	.35	0.3
Tensile strength (long), MPa	1.75	438
Fracture Toughness, kPa $\sqrt{m}$	709	
Coeff. of thermal expansion, /deg. C	$2.0(10^{-6})$	$2.44(10^{-6})$

For reflective cracking, the largest stress in the fiberglass overlay is generated with the B777 gear load in the position shown in Figure 13.2 on one side of the transverse joint.

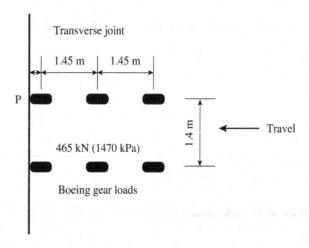

Figure13.2 B777 gear load on one side of the transverse joint

13.2 Stress analysis for an ACF overlay

Because of the relative difference in the flexural rigidities of the ACF and the PCC, the ACF contributes a negligible amount to the deflection of the PCC near the joint, so that the deflections are determined for the PCC pavement without the contribution of the overlay. The maximum bending and shear stress in the overlay are generated with the B777 gear load entirely on one side of the transverse joint is shown in Figure 13.2. Employing the method of stress analysis ***EFM*** described in Chapter 3, the deflection at any point Q with the load at P is

$$w_{QP} = \frac{2\left(1 - \upsilon^2\right)H}{PE} \int_0^l K_{1P} K_{1Q} \, dc \qquad (13.1)$$

where $K_{1P} K_{1Q}$ = the SIFs for the load at P and Q, respectively. P is the point on the transverse joint (Figure 13.3) and Q is the point under the wheel load for which the deflection is required. For a crack of length $2c$, in the PCC subjected to a maximum bending stress f_b, perpendicular to the joint, in the unjointed concrete pavement, the SIF is

$$K_I = \frac{f_b \sqrt{\pi c}}{1 + 0.125 \lambda^2 c^2} \qquad (13.2)$$

For a maximum shear stress τ, by analogy

$$K_{III} = \frac{\tau \sqrt{\pi c}}{1 + 0.125 \lambda^2 c^2} \qquad (13.3)$$

From the computer program **DEFCRACK,** the entire deflection profile of the concrete on both sides of the transverse joint caused by the main gear loads, obtained by using the computer program **EFM.** The FG overlay spans the distance b and the deflections, and slopes in the concrete and the fiberglass at points A and B, the points of tangency, are assumed to separately equal. The stresses and deflection in the concrete and the FG at A and B are obtained by structural analysis. The maximum bending and shear stresses in the fiberglass were found to be at A. The final deflection profile of the PCC at the joint has a fault of 0.070 cm as shown in Figure 13.3.

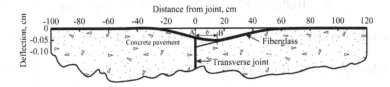

Figure 13.3. Deflection profile of the fiberglass overlay (grossly exaggerated vertical axis) with the B777 main gear load entirely on one side of the transverse joint

Table 13.1 Material properties of the AC, PCC and Fiberglass

Material	AC	PCC	FG
Modulus, MPa	3450	29,100	17238
Bending strength, MPa			438
Fracture toughness	500	905	
Expansion coefficient mm / mm / deg C		$1.0 \, (10^{-5})$	$2.44 \, (10^{-5})$

The maximum bending stress $f_b = 41.3 \, \text{MPa} \, (5.98 \, \text{ksi})$. The maximum shear stress 5.5 MPa *(7.99 ksi)*, so that the maximum shear stress = 3/2 (5.5) = 8.25 MPa (1198 psi). The combined stress for mixed mode fatigue failure is approximately $\sqrt{41.3^2 + 5.5^2} = 41.6$ MPa

13.3 Fatigue of the ACF overlay

Where there is a full bond between the ACF overlay and the PCC, the thermal stress is very small. But over the unbonded length of 30 cm the FG must carry the induced strain during contraction of the concrete. Assuming a net relative stretch between the concrete and the FG overlay of $\delta = 0.2$ cm, the tensile stress is

$$\sigma_t = E^* \frac{\delta}{l} \qquad (13.5)$$

where $l = 30$ cm. Assuming a long term FG modulus of 0.5 E^* (Table 13.1), the thermal stress as 113 MPa (16.4 ksi). This value is small compared to the tensile strength of 450 MPa. Experimental fatigue data on FG by Creed (1993) show that the fatigue limit, based on 10 million cycles is 20% tensile strength or 90 MPa. The combined bending and shear stress is 40 MPa. Consequently the fatigue life of the FG overlay is likely to exceed 100 million cycles by conservative extrapolation of the data. As pointed out previously the thermal stress does not add algebraically to the load stresses, but merely increases the minimum value of the SIF generated by the loads. Thermal fatigue by itself has a frequency of only once per day, so that thermal fatigue per se is insignificant.

Joint efficiency is defined as the ratio of deflection of the unloaded and the loaded sides of the PCC slab. The deflection profile of the PCC slab caused by the joint alone under a heavy gear load entirely on one side is shown in Figure 13.3. The joint efficiency in typical PCC pavement design is about 0.88. Lower joint efficiency are caused by dowel bar looseness, loss of aggregate interlock, cracking of the CTB which lead to debonding and loss of contact of the CTB from the PCC and water entering through the joint, causing softening of the subgrade, pumping and faulting at the joints. However, the joint efficiency of the FG load transfer at the joint is likely to be over 97%, because water cannot enter the opening in the concrete pavement.

13.4. Life of the AC conventional versus that of the ACF overlay

The computer program called **REFLECTION** determines the stresses in various sections of the ACF overlay and conventional AC overlay, in accordance with the foregoing equations. The vehicular and thermal stresses are given in Table 13.2.

Table 13.2 Stresses in the AC and ACF Overlays

Overlay	Vehicular Stresses (MPa)		Thermal stress (MPa)
	Bending	Shear	Tensile
E-glass polyester comp.			
@ joint (15 mm thick)	39.7	8.25	113
E-glass polyester @ midspan			
@ midspan (1.9 mm thick)	1.46	0.35	51.91
AC @ midspan (2.35 mm thick)	-0.75	0	0.36
Conventional AC (30.5 cm thick)	0.51	0.72	0.34

As an alternative to the ACF overlay, a conventional AC overlay 30.5 cm thick may be used. It was assumed that the annual temperature fluctuates 22 C sinusoidally with a gradient of 0.65 C/cm, and the annual traffic was taken as 2000 equivalent Boeing 777 gear loads. The number of cycles for the reflective crack to appear in the surface of the 30.5 cm overlay was

59000 cycles. The number of cycles for reflective cracking as a function of the thickness of the overlay is presented in Figure 13.4.

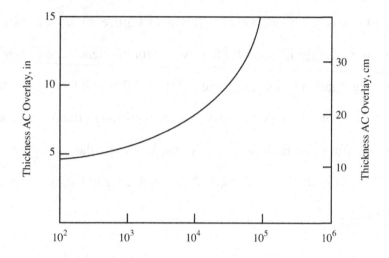

Figure 13.4.Fatigue life of a conventional AC overlay

13.5 Construction of an ACF overlay

Since this type of construction has not been done before, new construction techniques are required. E-glass mat reinforced polyester sheets 2.5 cm thick are manufactured by some companies, such as Fluid Containment Inc. They are transported to the airfield and placed over a bonding coat of polyester on the concrete. Since polyester is about 100 times stronger than AC, there will be no loose stones to damage the jet engines of the aircraft in service. Alternately the top surface of the FG can be made skid resistant by texturing during manufacture at the factory. Each manufactured sheet should have a projection of at least 15 cm of the fiber reinforcement at each end so that 15 cm gap can be filled during construction by polyester to form a continuous overlay.

The remainder of the overlay consists of 0.15 cm thick fiberglass pre-manufactured and bonded to the concrete with polyester resin. A tack coat of AC-20 is then applied to the fiberglass which is then overlaid with 2.35 cm of AC.

13.7 Approximate cost analysis of the ACF overlay

The cost of the new overlay $(2.5 \, m^2)$ including construction costs for material cost + labor are:

E-glass mat reinforcement polyester	$35.00
25.4 mm AC	$6.00
Total cost for new overlay	$41.00
Cost of conventional overlay	$80.00

If there are interior cracks in the PCC the thickness of the ACF can be increased from 0.19 cm to the appropriate value determined from *EFM*. The estimated cost of the ACF overlay is less than the cost of a conventional AC overlay. The ACF overlay can withstand the thermal stresses and one million cycles of loading from Boeing 777 aircraft, whereas the AC overlay will crack in about 59000 cycles. The computer program is user-friendly and based on fundamental principles. All modes of crack are included. The proposed innovation captures the fundamental behavior of reflective cracking in overlays and eliminates reflective cracking.

PROBLEMS

13.1

Consider a pavement with the following properties: AC surface: $E_1^* = 1400$ MPa, $H_1 = 10$ cm. Concrete, 28-day strength $= 34.5$ GPa, $H_2 = 15$ cm. Silty clay subgrade: 62 MPa. The AC surface has typical alligator cracking with the average longitudinal crack 40 m and transverse crack 10 m. The design traffic consists of 500,000 cycles of 90 kN ESALs over a period of 10 years. Determine the life for reflective cracking to occur.

14

RATIONAL STRUCTURAL DESIGN
OF HIGHWAY PAVEMENTS

"When theory and practice blend, the result is exciting and formidable"

When there are rational solutions, the art and science of design becomes exciting. Not only is it simpler, but it inspires the engineer to attain the highest quality product, with engineering efficiency and business economy.

14.1 Flexible Pavements

For flexible pavements fatigue in bending of the surface course is the dominant feature. All suitable materials should be tested and the associated costs of the materials delivered to the site should be estimated. Certain minimum requirements should be met. For the subgrade, the undrained shear strength and the compressibility are required. Usually the absolute minimum undrained shear strength of 5 psi and a minimum dry density of 100 pcf are required. For a suitable subbase, the permeability should be at least 100 ft./day (beach sand) with a modulus of at least 20,000 psi. For the base course the dynamic modulus E^* should be at least 40,000 psi.

14.2 Concrete Pavements

Plain concrete pavements can be built on almost any type of subgrade. The subgrade must be uniform and free of sulphates that attack concrete chemically and organic material should be less than that specified by ASTM ignition test. All non-uniformities should be excavated and replaced by compacted fill, or treated to make them satisfactory. Typical joint spacing is about 15-19 ft. (4.6 m). If the spacing of the joints is greater than 19 ft. two transverse cracks may develop simultaneously. According to CALTRANS, the most prevalent problem in PCC pavements is faulting at the joints. For the design of dowelled joints the reader is referred to their respective State DOT manuals.

The primary design is against the fatigue mode of distress. The elastic modulus of PCC is at least six to eight times greater in compression than in tension. The bending modulus, is given as

$$E_b = \frac{2E_t}{1 + \sqrt{\dfrac{E_t}{E_c}}}$$

(15.2)

where E_c and E_t = moduli of concrete in compression and tension, respectively. Taking $E_c = 6E_t$, gives the bending modulus $E_b = 0.24E_c$. The dominant fatigue mode is bending in Mode I, with crack propagation rate, as stated in Eq. (6.19) as:

$$\frac{dc}{dN} = 0.153 \frac{(\Delta K - \Delta K_{th})^2}{E_b^* f_r \left(1 - K_n^2\right)}$$

(15.3)

The maximum tensile stress at the bottom of the pavement with the wheel load tangential to the longitudinal edge as described by the Westergaard (1929) is used for concrete pavement design. Some design procedures use the maximum bending stress generated by the aircraft in the interior of the pavement (Packard, 1973). The width between the centers of the gear loads for a Boeing 777 aircraft is 10.97 m. For a runway joint spacing of 7.62 m, the aircraft, if centered on the runway with respect to the wheels and the joints in the PCC, would travel with edge of the wheel load 1.21 m from the nearest longitudinal joint. The actual traffic pattern has a wander of ∓ 9.0 m (Packard, 1973). From the distribution of the traffic approximately 7.5% of the gear loads fall in the vicinity of the longitudinal joint. The bending stresses here are considerably greater than those at the interior. The Federal Aviation Agency (FAA) Circular QAC150-5320-6C recommends the use of 75% of the *Westergaard edge bending stress.* This was obtained by correlation with the elastic interior stresses with the edge stresses for different aircraft types and subgrade strengths.

A major shortcoming in the current empirical design procedure is the manner in which thermal curl stresses are handled. If added to the aircraft stress, the final stresses would be so large, necessitating the use of calibration factors (*in excess of 10 times*) to bring the results into line with field observations. Such large calibration factors raise the question about the validity of entire procedure. According to the principles of fracture mechanics, the thermal curl stress increase the minimum stress, but it is the amplitude of the cyclic stress generated by the aircraft loading that is primarily responsible for fatigue cracking.

14.3 EVAPAVE

This is a new material which the author invented from the theories presented herein for highway and airport pavement design. For fatigue properties, search was made for *high tensile strength, high fracture toughness, a bonding ingredient and the molecular structure for cross linking.* The tensile strength was of paramount importance. The new material turned out to be a copolymer, ETHYLENE VINYL ACETATE (EVA) with the properties given in Chapter 10.

14.4 Computer programs used in the design of the pavements

The designs are considerably facilitated by the use of the following computer programs which are in a CD at the end of the book.

Table 14.1 List of computer programs used for design and analysis

Pavement type	Stresses	Fatigue	Rut depth
Flexible	***CHEVRON***	***ALLIGATOR***	***RUT*** (AC)
	EFM		***SUBGRADE (Sub.)***
Concrete	***WESTRESS***		
	CHEVRON + EFM		
EVAPAVE	***CHEVRON + EFM***	***ALLIGATOReva***	

14.5 Fatigue cracking and rutting in pavements

For viscoelastic/plastic materials, such as *concrete, asphalt concrete, EVAPAVE and other composites* the rate of crack propagation per cycle is as given in Chapter **7**. The rut depth in subgrades was presented in Section 8 and computerized in *SUBGRADE* and that for AC in *RUT*. All elements of rational design are in place, except for the endurance or fatigue limit, which follows.

14.6 The endurance limit or threshold SIF

The endurance limit is not fully understood. There is no theoretical or empirical expression for the values for metals, concrete, asphalt concrete or EVAPAVE. The following information for metals and concrete obtained by experimental testing is given in Table 14.1.

Table 14.1 Endurance of fatigue limit for various engineering materials

Material	Fatigue limit	Comments
Steel	$0.5\,f_{ult}$	well established by testing
Al, iron, and copper	$0.4\,f_{ult}$	well established by testing
P. C. Concrete	$0.5\,f$	Murdoch (1959), Raithby and Galloway (1974), Ballinger !970),
Asphalt concrete	60	Thompson and Carpenter (2006)
EVAPAVE	$90\,E_b 10^{-6}$	estimate by Ramsamooj (2012)

The threshold SIF ΔK_{th} is obtained from the fatigue limit as described in the computer programs.

14.7 Design procedure

The design procedure will now be illustrated by examples.

14.7.1 Flexible pavement design example

Design a pavement to carry 5 million cycles of 18 k single axle (ESALs) loads in 20 years. The seasonal temperature variation is sinusoidal. The properties of the available materials, including cost of construction are given in Table 14.1. The subgrade consists of clay.

Table 14.2 Available material properties and estimated costs

Material	E_c psi	Ten. Str. f_b psi	c,∅ psi (deg.)	Perm. k k, ft./day	Cost $/cu. yd.	Cost/ f_b
AC I	282 000	319		0.22	50	0.156
AC II	329,000	385		0.12	55	0.143
AC III	162,000	365		0.12	62	0.170
AC IV	250,000	500		0.12	67	0.134
Sandy Gravel	40,000	0	0 ,46	500	25	-
Bit. Stab. Gravel	60,000	100	100,46	200	40	0.18
Coarse sand	30,000	0	0 ,46	100	20	-
Crushed stone	60,000	0	0, 60	100	22	-
Clayey gravel	35,000	5	10, 40	40	18	0.94
Sandclay	35,000	5	24, 35	40	15	0.20

Unit conversion: 1 psi = 6.895 kPa; ft./day = 30.5 mm/day; 1 cu yd. = 0.028 cubic meter

Surface layer

On the basis of cost/tensile strength shown in Table 14.1, the choice of the AC for the top layer is AC IV; the lower modulus will give a lower bending stress and a higher fatigue life. However, the modulus is too low so that enough protection to the lower layers is not provided; hence *AC II is the preferred surface course.*

Base

The modulus of the base course is generally expected to be smaller than the top layer by a factor of about 2. Gravels and sand cannot carry tension, but a small surface tension of about 1 to 2 psi is allowable, because of stress redistribution. Silty and clayey gravels can carry tension without cracking by fatigue if the tensile stress is smaller than the tensile strength by a factor of safety F_s of about 5. The relative shear strength of crushed stone, gravel or clayey gravel may be evaluated as being proportional $\tau_f = c + \sigma_n \tan\phi$, where the normal stress is taken as roughly 10 psi for the base course and 7.5 psi for the subbase.

Table 14.1 shows the shear strength parameters and thecost/cu. yd./tensile strength. Accordingly, the most economical material *is bituminous stabilized sandy gravel.* For a first class highway, 6 in. of AC surface is good practice. Since *bituminous stabilized gravel* is more economical than sandclay, try 8 in. of this material for the base.

Subbase

The minimum thickness of any layer is usually 3 inches because of construction. The subbase is required to provide most of the drainage as is usually the case, but may be required for protection of the subgrade. Try 6 in. thick *silty gravel* for the subbase as shown in Table 15.2.

Trial design

Table 14.2 Thickness and strength of the paving materials

Layer	Thickness, cm.	Bend. Modulus, MPa	Tensile/Shear str., kPa, deg.		v
AC surface	15	1000		2379	0.35
Bit.-St. base	20	414		1241	0.35
Silty gravel subbase	15	138		55	0.30
Clay subgrade			62	69, 25 (c, ϕ)	0.45

Stresses from **CHEVRON**

For each layer we need the bending stress at the bottom of each layer for fatigue and the principal stresses and the vertical displacement after the first cycle for rutting. The max shear stress at the edge of the loaded area may be needed for Mode III type of failure (TDC), depending on its magnitude relative to the Mode I fatigue bending stress. These properties are shown in Table 14.3. Design stresses (tension +) are highlighted in bold.

AC surface: Primary fatigue   shear stress Modes I + III $\tau_{max} = 200 +$ **bend f_b. 16.54** kPa

Mode I + mode III tearing fatigue eq. Mode I			$f_{b,} = 231\,\text{kPa}$
Bit. Stab. base: Primary fatigue		bending stress	$f_b = 71.4$ kPa, Mode I
Silty gravel	bending stress	tensile stress	$f_t = 26.6$ kPa, Mode I
Clay subgrade		shear stress	$\tau_{max} = -13.5$ kPa $\left(\text{check B.C}\right)$

The stresses in the Table 14.2 appear to be reasonable, so accept it as a trial design. It is then necessary to check the fatigue life of the AC, BS base, subbase and the rutting of each layer as a function of the number of cycles of loading.

Fatigue—AC surface

Since the shear stress at $R = 15$ cm, $z= 5$ cm $\tau = 200$ kPa exceeds the bending stress = 192 kPa at $R= 0$, $z= 15$ cm, Mode III or the tearing mode will be the primary fatigue mode of distress. But there is also a bending stress of 114 kPa at $R= 15$, $z = 15$ cm. Therefore the secondary mode is Mode I, the opening mode. The equivalent mode $K_{Ieq.}$ may be calculated by Eq. (6.16) as

$$f_{beq} = \sqrt{(16.54)^2 + 0.866(29.09)^2} = 252 \text{ kPa}$$

Table 14.3 Stresses and material properties

Radius in.	Depth. in.	Layer	Max. $f_{b,}$ psi	Max. τ, psi psi	Fat. Mode	Design stress psi
0	6	AC	27.72	0		27.72
6	2	AC	-47.60	**-29.09**	I + III	**33.47**
6	6	AC	**16.54**	-14.12	I	27.72
0	8	BS Base	0.11	**10.36**		**10.36**
0	14	BS Base.	**16.58**	0	I	16.58
0	20	Silty-gravel	**3.85**	0		**3.85**
0	22	Clay subgrade		**-1.9**		**-1.9**

Unit conversion: 1 psi = 6.895 kPa; 1 in. = 2.54 cm

AC surface layer (ALLIGATOR)

No. of cycles for the primary longitudinal crack to grow to 155 cm = **1.41 million**

No. of cycles for the transverse crack to grow to 61 cm = **2.69 million**

Bituminous stabilized base

No. of cycles for the secondary transverse and longitudinal cracks to grow to 166 cm = 0.8 million cycles.

Therefore the fatigue life of the pavement is limited by the lack of strength and thickness of the AC surface and more by the bitumen-treated base. Revise the design as shown in Table 14.4. The revised stresses are given in Table 14.5.

Table 14.4 Revised design-thickness and strength of the paving materials

Layer	Thickness, in.	Bend. Modulus, MPa	Ten./Shear strength., kPa	v
AC surface	15	1310	3034	0.35
Bit.-St. base	25	414	1241	0.35
Silty gravel sub.	15	137	55	0.30
Clay subgrade		63	69, 25 (c, ϕ)	0.45

Table 14.5 Revised design stresses

Radius in.	Depth. in.	Layer	Max. f_b, psi	Max. τ, psi	Fat. Mode	Design stress psi
0	8	AC	**36.53**	0		**36.53**
8	2	AC	-48.60	**-14.34**		
8	8	AC	**10.33**	-14.12	I	-
0	18	Bit. St. Base.	**10.33**	0	I	**10.33**
0	20	Silty-gravel	**1.76**	0		**1.76**
0	22	Clay subgrade		**-1.19**		**-1.19**

Unit conversion: 1 psi = 6.895 kPa, 1 in. = 2.54 cm

The completed design is presented, plan and cross section in Figure 14.1a and 14.1b.

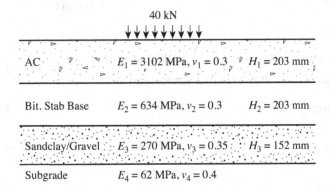

Figure 14.1a Design of the flexible pavement

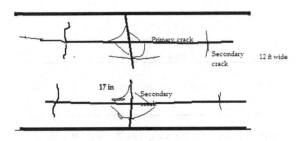

Figure 1 Flexible Pavement Crack Pattern

14.7.2 Concrete pavement design example

Design a concrete pavement for a fatigue life of 20 million 80 kN ESALs for a design period of 20 years. The available materials include cement stabilized bases, sandy gravel, and sand. The subgrade is uniform sandy clay.

Solution

This is a relatively simple choice of base as sandy gravel. The uniformity of the subgrade is an essential property, since uncontrollable settlement quickly leads to failure for concrete pavements. The material properties are:

Concrete: 28-day strength	$f_c^{"} = 34.48$ MPa
Modulus in compression	$E_c = 27.80$ MPa
Modulus in tension	$E_t = 6.67$ MPa

CONPAVE Input data

Maximum bending stress with 9000 lb. wheel load tangent to longitudinal .edge from Westergaard stress analysis	$f_b = 235$ psi
Modulus of rupture	$f_r = 707$ psi
Fracture toughness	$K_{1c} = 1022$ psi $\sqrt{in}$
Bending modulus	$E_b = 967,200$ psi
Fatigue endurance limit	$= f_{el} = 353$ psi

40 kN

PCC $E_1 = 3102$ MPa, $v_1 = 0.3$ $H_1 = 203$ mm

Sand $E_2 = 206$ MPa, $v_2 = 0.3$ $H_2 = 203$ mm

Subgrade $E_3 = 62$ MPa, $v_3 = 0.4$

Figure 14.2(a). PC Concrete Pavement

194

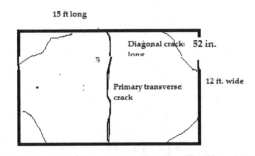

Figure 14.2(b) Concrete Pavement Crack Pattern

CONPAVE **Output**

Inverse characteristic length $\qquad \lambda = \sqrt[4]{k/D} = 0.051$

Flexural rigidity $\qquad D = EH^3 / \left(12\left(1 - v^2\right)\right)$

Distance from corner *along the corner angle bisector* = 25.9 inches (*l*= 52 in)

Number of ESALS (N_1) for **the primary crack** to traverse the width of the pavement lane (438,000 cycles for crack initiation)

$$N_1 = 10.98 \text{ million cycles}$$

Number of ESALS (N_2) for **the secondary cracks** comprising the four diagonal cracks approx. 52 in. long to develop as shown in Figure 3 (630 days for crack initiation or 630, 000 cycles).

$$N_2 \gg 14.43 \text{ milion cycles}$$

Therefore the fatigue life of the 20 cm. thick pavement is 14.43 million cycle of 18-K ESALS at a rate of 1 million per year.

The fatigue analysis does not include any other form of distress, such as transverse joint faulting, which is the most prevalent form in California.

14.7.3 EVAPAVE design example

Preliminary design

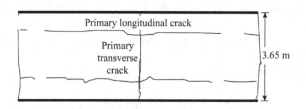

40 kN (552 kPa)

EVAPAVE surface	$E_{1b}^* = 8274$ MPa	$v_1 = 0.25$	$H_1 = 20$ cm
Low quality EVAPAVE base	$E_{2b}^* = 1723$ MPa	$v_2 = 0.25$	$H_2 = 20$ cm
Sand	$E_3 = 207$ MPa	$v_3 = 0.35$	$H_3 = 15.2$ cm
Subgrade	$E_4 = 61.7$ MPa	$v_4 = 0.4$	

Figure 14.3a Cross section of EVAPAVE

Primary longitudinal crack

Primary transverse crack

3.65 m

Figure 14.3b Crack pattern in EVAPAVE preliminary design

ALLIGATOR **input data @ 25 deg. C**

Maximum bending stress with 9000 lb. in wheel path. $f_b = 61.8$ psi

Indirect tensile strength $f_{ct} = 720$ psi

Fracture toughness $K_{Ic} = 1425$ psi $\sqrt{\text{in}}$

Bending modulus $E_b = 1,200,000$ psi

Endurance limit $= 90 E_b 10^{-6}$

ALLIGATOR Output

Inverse characteristic length $\qquad\qquad\qquad \lambda = \sqrt[4]{k/D} = 0.080$

Flexural rigidity $\qquad\qquad\qquad\qquad\qquad D = EH^3 / \left(12\left(1 - v^2\right)\right)$

Distance of diagonal cracks from corner *along the corner angle bisector* = 22 inches

Number of ESALS (N_1) for the primary crack to traverse the width of the pavement lane, @ 1 million cycles per year

$$N_1 = 56.5 \text{ milion cycles}$$

Number of ESALS (N_2) for the secondary cracks comprising the four diagonal cracks approx. 4 ft. long, , @ 1 million cycles per year

$$N_2 \gg 100 \text{ milion cycles}$$

Therefore the fatigue life of the preliminary EVAPAVE design is practically unlimited.

The primary cracks, consisting of transverse cracks occur at a spacing of about 15 ft.(Figure 14.3) At failure when the transverse cracks coalesce across the full width of the 12 ft. lane,

$$\lambda c = 2.88$$

which is close to the arbitrarily defined normalized crack length $\lambda c > 3.5$ However, the secondary cracks which should occur along the diagonal at the intersection of the primary cracks do not initiate even under a 50% overload axle loads which is about 0.36% of the total load distribution spectrum according to FHWA, plus the thermal stress plus 15% for impact at the intersection of the primary cracks. Only vehicular loads and daily cycles of thermal stresses are considered in the computations. The effect of seepage of water through cracks is not considered.

14.7.4 EVAPAVE final design optimization

It is proposed to improve on the above preliminary design by fine tuning. The goal is to obtain a design in which the endurance limit is not exceeded in any layer, so as to guarantee long fatigue life. The design stresses for the above pavement is presented in Tables 14.6a and 14.6b for the preliminary and final designs. Only fatigue cracking is considered, since EVAPAVE shows negligible rut.

Table 14.6a Stresses in pavement—preliminary design –

Layer	H cm	f_b, kPa	τ, kPa	$1.4 f_b / f_{end}^{*}$	Comment
Surface (EVP)	20	426	202	0.82	no crack initiation
Base (LQEVP)	20	136	49.5	1.22	crack initiation
Sand subbase	15	14	36.4	-	

$$f_{end} = 90 E_b (10)^{-6}$$

Table 14.6b Stresses in the pavement – final design

	H cm	f_b, kPa	τ, kPa	$1.4 f_b / f_{end}^{*}$	Comment
Surface (EVP)	15	315	202	0.59	
Base (LQEVP)	23	191		0.97	
Subbase, gravel	15	14			
Subgrade		0. 6			

Tables 14.6 and 14.7 show that the bending stress is dominant, so that the design criterion is bending fatigue. An overload factor of 1.4, corresponding to an axle load of 1.4(90) = 126 kN is used to determine if cracks will initiate. From Table 14.6, the preliminary design pavement will have crack initiation under 126 kN axle loads, but in the final design cracks will not initiate. **Accordingly the pavement design shown in Figure 14.3a with a top surface layer of 15 cm and a base course of 22.5 cm will have an unlimited fatigue life. The optimization produces lower cost and a superior pavement.**

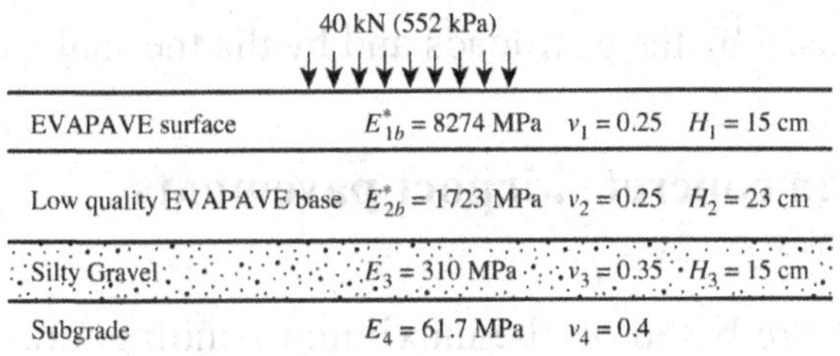

40 kN (552 kPa)

EVAPAVE surface	$E_{1b}^{*} = 8274$ MPa	$v_1 = 0.25$	$H_1 = 15$ cm
Low quality EVAPAVE base	$E_{2b}^{*} = 1723$ MPa	$v_2 = 0.25$	$H_2 = 23$ cm
Silty Gravel	$E_3 = 310$ MPa	$v_3 = 0.35$	$H_3 = 15$ cm
Subgrade	$E_4 = 61.7$ MPa	$v_4 = 0.4$	

Figure 14.4 Cross section of the final design of an EVAPAVE pavement

14.8 Advantages of EVAPAVE

1. No joints or cracks or ruts

2. Smoothest driving pavement

3. Use of recycled EVA enhances the environment

4. More economical than either asphalt concrete or concrete

5. Large savings in fuel cost and maintenance cost of highways, airports pavements, cars, trucks and airplanes

6. Will outlast any other pavement by over seven times

15

RATIONAL STRUCTURAL DESIGN
OF AIRPORT PAVEMENTS

A rational method of thickness design of plain concrete runways and taxiways against the fatigue mode of distress is presented. The material properties used for design are the tensile strength, the fracture toughness, the Young's modulus, Poisson's ratio and the endurance limit. The new method of stress analysis comprising the use of multilayered ***Elastic theory combined with Fracture Mechanics (EFM)*** (**Chapter 4**) is used to determine the stresses in the jointed pavements caused by the gear loads and by the thermal curl stresses.

15.1 Design stresses in concrete airport pavements

Some design procedures are based on the maximum bending stress in the interior of the pavement generated by the aircraft load (Packard, 1973). From the actual distribution of the traffic, approximately 7.5% of the gear loads fall in the vicinity of the longitudinal joint, so that the bending stresses are considerably greater than those at the interior. The Federal Aviation Agency (FAA) (FAA advisory Circular AC150-5320-6C uses 75% of the Westergaard edge bending stress for the design stress.

A trial design for an airport pavement is presented in Figure 15.1a, and the location of the gear loads for maximum bending stress in the pavement is shown in Figure 15.1b. The maximum stress usually occurs at the bottom of the pavement in the location shown in Figure 15.1b with the main gear load tangent to the longitudinal edge. ***EFM*** is used to calculate the maximum bending stress for any aircraft. It shows that if the width of the outer lane is made

30 in (75 cm) wider to avoid the effect of increasing the stress at a point 36 in (91 cm) inside of the edge, the design stresses may then be determined by using **CHEVRON.** We may either use 75% of the edge stress for design *(Packard, 1973),* or alternatively, find the customary landing distance x_i from the longitudinal edge for the aircraft of the design spectrum of aircraft, and compute the **design** wheel load position of the "reference" wheel load P', where the ' denotes the part of the wheel load used to overcome the endurance limit' or

$$x = \frac{\sum_1^n x_i P_i'}{nP'} \qquad ()$$

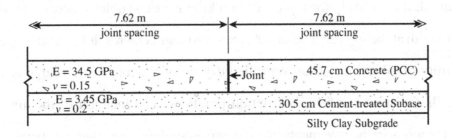

Figure 15.1a Cross section of the airport pavement

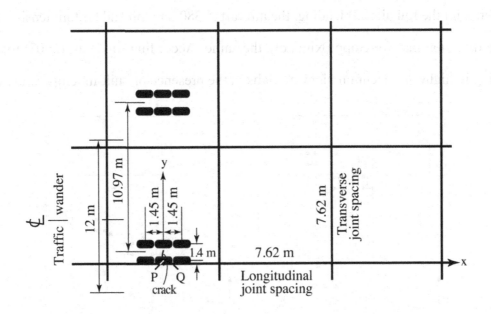

Figure 15.1b Location of B777 gear loads for maximum

bending stress at the bottom of the concrete layer.

Recent observations from full-scale rigid airport at FAA's National Airport Pavement Test Facility and Airbus Pavement Experimental programs have shown that top-down cracking can occur (Roessler *et al.* (2007). Guo (2006) observed that top-down cracking can occur even for thick concrete slabs in various locations caused by the gear load at both the longitudinal and transverse joints. Finite element analysis indicates that all of the aircraft gear loads are necessary to obtain accurate top tensile stresses. Roessler *et al.* (2007) determined from finite element analyses the B-777 gear load produced the highest ratio of the top and bottom tensile stresses of 0.50. They pointed out that when temperature curl is omitted, it is unlikely for top-down cracking to occur unless there is a reduction of concrete strength at the top concrete of 50%. Such strength can result from paving efficiency, nonuniformity of vibration and uncontrolled drying of the top. Such a large reduction in strength calls for more careful construction practices. The location of the aircraft loads relative to the joints in the pavements and the corresponding maximum top bending stresses are presented graphically in Figures 15.5a and b and in Figures 15.6a and b. For no joint load transfer, with the full aircraft loading, the aircraft A-380 had top and bottom tensile stresses in one direction that were approximately the same. According to Heath (2003) top-down cracking in highways occur on highway slabs in the presence of built-in temperature curl.

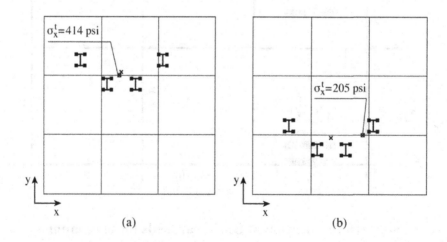

Figure 15.5 Location of gear loads for maximum top-down bending stress for Aircraft B777 (Roessler, 2007)

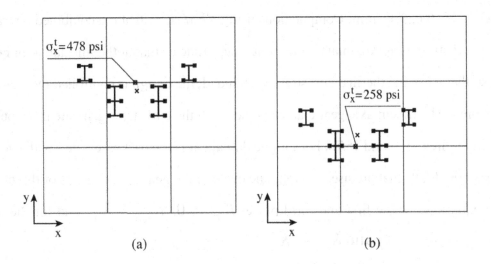

Figure 15.6 Location of gear loads for maximum top-down

bending stress for Aircraft B 777 (Roessler, 2007)

15.2 Fatigue cracking in concrete airport pavements

There is a seasonal variation of the thermal gradient, which can be obtained by extrapolating the results of the AASHO Road Test for a 43.9 cm thick concrete pavement (Darter and Barenburg, 1977). The daytime temperature gradient is

$$\theta = 0.0124 + 0.0096\left(1 + \frac{2\pi N}{2500}\right) \tag{15.1}$$

where N = number of applications of the tridem gear load. The night-time thermal gradient is assumed to be constant at 0.11 C/mm. When the curled slab is supported at the corners and the weight then exerts a stress that is determined as follows. Using the *EFM* method, the thermal curl stresses are obtained as described in **Chapter 11.8** and can be obtained for either an upward or downward curl.

15.3 Stresses generated by the tridem axle gear loads, Boeing 777, airfield pavements

The maximum gear load stresses occur at the longitudinal(x) joint midway between the transverse joints for the loading shown in Figure 15.7b. The maximum thermal gradient was assumed to be $0.022^0 C/mm$. Computations using **EFM** show that the combined aircraft and thermal curl stress is approximately the same magnitude as that at the midslab joint bending stress, so that when the thermal curl stress is included, the design is controlled by the midslab joint loading. The tridem axle gear loads together with the curls tress generate three pulses as shown in Figures 15.7a and 15.7b. For fatigue damage it is the ΔKs associated with each peak or the rises and falls that are used, so that one cycle of the gear load consists of three . In the fatigue equation, for the first and third pulses $K_{min} = 0$, $K_{max} = K_a$, and for the second pulse $K_{min} = K_{min} = K_b$ and $K_{max} = K_c$.

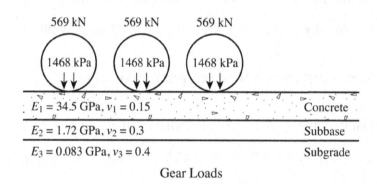

Figure 15.7a B777 gear loads on concrete airport pavement

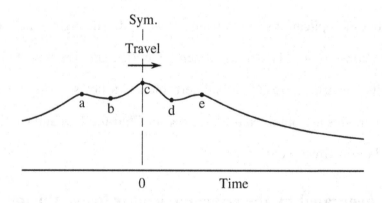

Figure 15.7b Pulses generated by B777 gear load

15.6 Computer program *CONPAVE* (App. VIII)

Any lateral distribution of the axle loads across the pavement can be used. A typical one is presented in Table 15.1.

Table 15.1 Distribution of gear loads in typical airports (Packard, 1973)

Position measured from longitudinal edge, m	Percentage of gear load
-4.27 to 0.25	7.5
0.25 to 0.5	8.5
0.5 to 0.89	9.0
0.89 to 7.92	75.0

CONPAVE input

- number of layers

- *b*ending modulus and Poisson's ratio of concrete

- compressive modulus and Poisson's ratio of subgrade

- thickness of concrete

- number of radii and radii where stresses and strains are required

- number of depths and depths where stresses and strains are required

- modulus of rupture

- maximum size of aggregate of concrete mix

- seasonal thermal gradient

- number of tridem axles per year

CONPAVE Output

- deflection and stresses at the midslab joint position

- the stresses

- the thermal curl stress

- the crack length as a function of the number of gear loads

The following pavements were analyzed.

Concrete

Thickness	= 406, 439, 508 mm
Modulus of elasticity E_c	= 34.5 GPa
Compressive strength	= 34,480 kPa
Modulus of rupture, f_r	= 62.05 (6.89) $\sqrt{f'_c}$ kPa
Endurance limit	= 0.50 f_r

Subbase-cement-treated

Thickness, mm	= 439
Modulus of elasticity, GPa	= 172
Compressive strength, kPa	= 4500

Clay subgrade

Modulus of elasticity, MPa	= 82

The maximum stresses generated by the gear load (75% of the longitudinal edge stress) and thermal curl for each pavement thickness is presented in Figure 15.8. The fatigue life computed by **CONPAVE** under the combined stresses from the gear loads and thermal curl, and from the gear loads alone, is presented graphically for three pavements in Figure 15.9. The fatigue life of the pavement with the combined effect of the gear loads plus the thermal curl stress is seen to be slightly smaller than that with the gear loads alone. From this graph, the design thickness for 500,000 applications of the aircraft is found to be 457 mm.

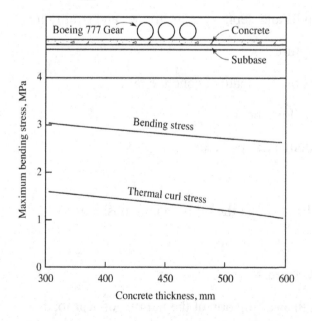

Figure 15.8 Maximum bending stresses vs. concrete thickness, B777 aircraft gear load

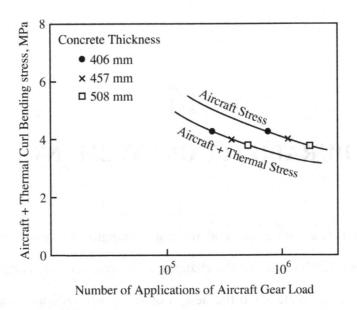

Figure 15.9 Aircraft stresses vs. the number of applications of the gear load

PROBLEMS

15.1

Design a rigid pavement 500,000 applications of a 500 kN B777 wheel loads for fatigue life of 20 years. The subgrade consists of a sandclay with a modulus of elasticity of 140 MPa.

15.2

Design an EVAPAVE airport pavement for 500,000 applications of a 500 kN B777 wheel loads for fatigue life of 20 years. The subgrade consists of a sandclay with a modulus of elasticity of 140 MPa. Use EVAPAVE with the properties given in Section 10 and gravel for the subbase.

16

DURABILITY OF PAVEMENTS

Asphalt cement hardens with aging and undergoes chemical changes. The task of the structural designer is to characterize the changes in the engineering material properties so that they could be incorporated into the design of AC pavements for fatigue, rutting and durability.

16.1 Effects of aging on the tensile strength, fracture toughness, and modulus

All of these parameters behave similarly with respect to aging. They increase with aging up to 5-8 years and then decreases asymptotically, becoming brittle. This may be expressed as

$$K_{Ic} \rightarrow K_{Ic}\left(1 + at + bt^2\right) \qquad (16.1)$$

where a and b are experimentally determined constants , and t = calendar time in years

16.2 Rate of hardening of asphalt cement pavements

Changes in the value of the rate of hardening $\lambda = dlnp / d\varepsilon_v^p$ changes logarithmically with age This may be expressed as

$$\lambda \rightarrow \lambda(1 + \gamma \log t) \qquad (16.2)$$

where γ is an experimentally determined constant, and t = calendar time in years.

16.3 Other durability factors

These include polymerization or chemical changes, degradation, stripping of the asphalt from the aggregates by moisture, disintegration from the lack of soundness of the aggregates, etc. abrasion resistance, stripping, and raveling are more difficult to evaluate quantitatively at the present time and should be estimated from laboratory tests.

The first two factors K_{Ic} and λ are directly incorporated into the equations for fatigue and rutting. The effect on performance of the pavement of the other factors may be estimated approximately.

17

DESIGN OF OVERLAYS

Before an overlay can be designed, it is necessary to evaluate the condition of the existing pavement. There are excellent manuals on this difficult task, such as government documents. The principal problems encountered are different for flexible and rigid pavements.

17.1 Definitions of Performance

Stability

Stability is defined as the ability to resist shoving and deformation under vehicular loads, so that the pavement can retain its smoothness without corrugations or other signs of shifting of the AC mixture.

Durability

Durability is defined as the ability to resist oxidation and polymerization (change of molecular structure), disintegration of the mixture and stripping of the binder from the aggregates.

Fatigue Cracking

Fatigue cracking is defined as the cracking that grows by blunting and resharpening of the crack tip under repeated vehicular loading eventually forming interconnected hexagonal cracks or *"alligator cracks"*, which is the main form of distress in most flexible pavements.

Skid resistance

Skid resistance is defined as the resistance of a surface against skidding or slipping of the vehicle tires, especially when wet.

Potholing is defined as the formation of holes that become progressively larger mostly in asphalt concrete pavements because of tire friction and impact of the wheel in hard and brittle surfaces that lack tensile strength.

7.2 Considerations for overlay construction on flexible pavements

A condition rating according to the existing *government manuals* should be performed, followed by test boreholes at a spacing of 50 to 150 m, depending on the repeatability. The following tests are essential:

1. The tensile strength of each layer, except for purely granular materials.

2. The shear strength of each material. For the subgrade the undrained triaxial test is recommended to obtain the shear strength parameters defining the shear strength envelopes.

3. A hydrostatic compression test for obtaining the parameter $\lambda = d\ln p / d\varepsilon_v^p$.

4. The density of each material.

Depending on the results of the condition survey and materials tests, the decision is made to construct an overlay or to breakup and reconstruct the entire pavement. Considerable judgment is required in making these decisions.

Usually if an overlay is chosen, the extent of cracking should not be so large that the load transfer capability of the existing AC is too low. The cracks and spacing may be considered to have "clusters" of alligator cracks, if the transverse and longitudinal cracks are spaced about 8 in. apart (or 20 cm hexagonal), with the transverse crack about 1.2 m long and the longitudinal crack about 4.1 m long. How to estimate to load transfer capability? This is discussed in the following.

The fundamental relationship between the total deflection of the crack-free pavement and the change in the deflection caused by cracks is given by Eq. 3.15 A computer program **DEFCRACK** computes for any crack spacing and length, the relationship between the cracking, corresponding change in the deflection of the pavement and the equivalent magnitude E^* of the AC *as if it were a continuum*. This enables **CHEVRON** to determine the deflection and stresses of the overlay for a specified traffic, design period and environment.

Consider the following scenario: the condition survey, including testing of the materials has been done. The results show that there are typical fatigue cracks comprising of clusters of transverse cracks 1.21 m long with longitudinal cracks 2.13 m long in the wheel path. There are also adjacent transverse cracks about 61 cm long with longitudinal cracks up to 183 cm long, roughly spaced 20 cm apart on both sides of the primary crack, and there may also be half-inch ruts in the wheel paths for about 40 % of the highway. There are no other signs of distress such as raveling, bleeding, corrugations, or pumping. The rut depths should be filled and the cracked area should be primed with bitumen with an interlayer placed over the cracked area. An AC overlay should then be designed and constructed over the pavement.

17.3 Overlay Design

Evaluate the capability of the present pavement to carry traffic. This requires the stiffness and strength of each layer to be evaluated for fatigue and rutting, which includes shear failure of the subgrade. For the cracked layer Eq. (3.28) is used to compute the increase deflection caused by the crack and the corresponding *equivalent dynamic modulus* E^*_{1eq} , which is defined as the dynamic modulus of the cracked pavement, treating it as a continuum. Except for the shear failure of the subgrade which is handled as part of rutting, the computer program **DEFCRACK** provides the answers.

Example 17.1

Design an AC overlay for the cracked pavement shown in Fig. 17.1

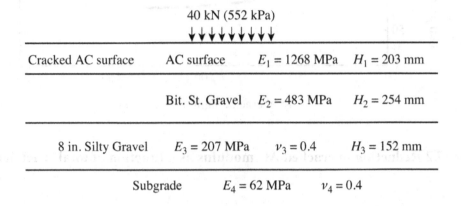

Fig. 17.1 Existing flexible pavement

Design Traffic

15 million 80 kN ESAL loads over a period of 15 years

Material properties *at start* of existing pavement with cracks

Strength

AC $f_b = 2655$ kPa ; Bit. St. sandy gravel (BSSG) : $f_b = 827$ kPa

Clayey gravel $c = 276$ kPa, $\phi = 40$ deg.:

Silty-Clay subgrade: c =20 kPa, $\phi = 30$ deg.

Recommended Procedure for Overlays

1. Determine the modulus of the cracked AC layer as if it were a continuous pavement slab, using fracture mechanics Use the computer program **DEFCRACK** to obtain the reduced magnitude of the complex modulus as a function of the crack length, transverse and longitudinal as shown in Fig.17.2. From this Figure the value of = 658 MPa.

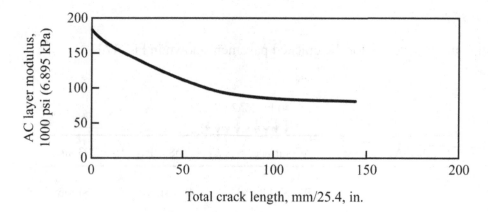

Figure 17.2 Reduction of cracked AC modulus as a function of total crack length

2. Determine the stresses in the cracked pavement

The stresses from **CHEVRON** are given in Table 17.1

Table 17.1 Stresses in the cracked pavement layers without the overlay

Depth, in.	Material	Modulus. MPa	Max. stress, kPa	Max f_b, kPa
-20	AC	655	98	2655
-45	Bit. Stab. Gravel	482	73	1393
-61	Clay gravel	207	49	-
-76	Clay subgrade	41	17	-

The stress analysis shows a tensile stress of 49 kPa at the bottom of the clay gravel layer, which is relatively high compared to the shear strength. A stiff overlay with a high tensile strength is needed. Use a high modulus, high tensile strength overlay, such as dynamic modulus $E_1^* = 3103\,\text{kPa}$, bending tensile strength = 3613 kPa.

3. Determine the stresses in pavement after cracking and with the overlay

The stresses from **EFM** are given in **Table 17.2**

Table 17.2 Bending stress at the base of the AC from *EFM*

Depth, in.	Material	Modulus MPa	Stress, kPa	Tensile str. f_b, psi
-15	ACO	1365 (bend)	262	3613
-36	AC (cracked)	659(bend)*	48	132
-61	Bit. Stab. Gravel	483	48	1393
-76	Clay gravel	207	18	-
-91	Clay subgrade	62	3.7**	-

* Reduced modulus as described above

**Clay subgrade: $Tension + ve : \sigma_1 = -9.4\,kPa, \sigma_3 = 0, \varepsilon_z = -0.000136$

The stresses appear reasonable so that this is **a good trial design**.

5. Check Fatigue Life of the new AC Overlay

Design period 20 years, 15 million ESALs

ALLIGATOR gives the following fatigue information on cracking:

ESALs for the transverse crack (spacing 40 cm apart) to grow to 61 cm and for the

Longitudinal crack to grow to 2(32) = 64 in. (total crack length = 2(24+32 = 112 in.) = 285 cm = 12.85 million ESALs for the transverse crack (spacing 81 cm apart) to grow to 61 cm and for the long .crack to grow to 2(34)= 68 in (total crack length = 112+2(24+34) = 579 cm = 14.94 million ESALs for the secondary transverse crack (spaced 81 cm apart) = 2.42 million ESALs for the transverse crack (spacing 71 cm apart) to grow to 86 cm and for the longitudinal crack to grow to 2(18)= 36 in (total crack length = 228+2(34 + 36) = 298 in.) = **27.6 million**

Therefore the overlay is estimated to improve the fatigue life satisfactorily.

6. Check the Fatigue Life of the Bituminous Stabilized Sandy Gravel

Fatigue life > 15 million after 20 years → *OK*

7. Check Rut Depth in each Layer

Input data to Computer Programs *RUT* **(AC, BIT. STAB. Gravel) and** *SUBGRADE* **(Subgrade soils)**

CHEVRON gives the stresses at the mid—depth of each layer and the vertical strain at 6 in. below the top of the subgrade. The permeability and coefficient of consolidation are obtained from soil tests in Table below.

Table 17.6 Material Properties of the Pavement

Layer	Depth.	Vert. stress	Hor. Stress	Vert. strain	Permeability	Consol. Coeff.
	cm	σ_1, kPa	σ_3, kPa	ϵ_z	k in/day	sq. in/day
AC overlay	-7.5	-47	-19	-0.000243	0.1	-
Old Cracked AC	-25	-15	-11	-0.000246	15	-
Bit. Stab. Gravel	-48	-15	24	-0.000123	6.0	
Silty Gravel	-69	-30	-0.42	0.000136	10000	200
Silty Clay	-36	-8	-0.001	0.00135	1.0	8

Shear strength: $c = 3\,psi, \phi = 30°$; $\lambda = 58, K = 3.0$; permeability $k = 2.5cm/day$, $c_v = 51.2\ cm^2/day$

Residual pore water pressure at the end of the first cycle = 4.4 kPa

Average vertical permanent strain for first few cycles $d\varepsilon_{y1}^p = 0.0005$

SUBGRADE computer output:

No. of cycles of 80 kN ESALs (250,000/yr.) = 2,500,000

Rut depth 0.76 cm. No shear failure.

RUTTING

See Problem 17.1

Bituminous Stabilized Gravel Base

Rut depth = 0.20 cm. after 15 million 80 kN ESALs

Silty Gravel Subbase

Rut depth = 0.36. after 15 million 80 kN ESALs

Silty Clay Subgrade

Rut depth = 1.68 cm after 15 million 80 kN ESALs

Total rut depth for the pavement with the 15 cm thick overlay = 224 cm

PROBLEMS

17.1

Consider a pavement with the following properties:

AC surface: $E_1^* = 1400$ MPa, $H_1 = 15$ cm

AC base $E_1^* = 1400$ MPa, $H_1 = 20$ cm

Silty clay subgrade: 62 MPa

The AC surface has typical alligator cracking with the average longitudinal crack 40 m and transverse crack 10 m. The traffic consists of 500,000 cycles of 80 kN ESALs. Design a suitable overlay for a period of 10 years.

17.2

For the pavement shown in Figure 17.1 calculate the rut depth in the overlay and the cracked AC base. You may make reasonable assumptions which should be completely stated.

18

RELIABILIY ANALYSIS

"Ultimately it is the reliability of the pavement design that must be ascertained before it is accepted for construction in practice".

18.1 Reliability Concepts

Ang (1975) presented the following reliability concepts. A performance function Z of the design variables is defined. For fatigue as described in Section 5

$$Z = N_f = g\left(\Delta K, \, K_{IC}, \, Y, \, E^*, \, c_0\right)$$

where ΔK = stress intensity factor, K_{Ic} = fracture toughness, Y = flexural tensile strength, E^* = magnitude of the complex modulus and c_0 = starter crack. Let μ_Z and σ_Z = mean and standard deviation of Z, so that

$$\mu_Z = N_f(\mu_{\Delta K}, \mu_{K_{Ic}}, \, \mu_Y, \, \mu_{E^*} \, \mu_{c_0}) \tag{18.1}$$

where μ = the mean value of the respective functions in the subscripts.

Then from a first order approximation

$$\sigma_Z^2 = c_i^2 \sigma_{Xi}^2 + \rho_{ij} c_i c_j \, \sigma_i \, \sigma_j \tag{18.2}$$

where

$$c_i = \frac{\partial N_f}{\partial \Delta K}, \qquad c_2 = \frac{\partial N_f}{\partial K_{Ic}}, \qquad c_3 = \frac{\partial N_f}{\partial Y}, \qquad c_4 = \frac{\partial N_f}{\partial c_0}$$

where $\sigma_{Xi} = \sigma_{\Delta K}, \sigma_{X2} = \sigma_{\Delta K}$, plus similar terms for Y, E and C_0 and ρ_{ij} is the correlation between the various parameters. Since there is none, $\rho_{ij} = 0$.

For example the total uncertainty in the value of $\Delta K = K_{max} - K_{min}$, including the mean stress range, Δf_b and the effects of impact and errors of stress analysis is

$$\Omega_{\Delta K} = \frac{\sigma_{\Delta K}}{\mu_{\Delta K}} = \sqrt{\left(\delta_{\Delta K}^2 + \Delta_{\Delta K}^2\right)} \qquad (18.3)$$

plus similar terms for K_{Ic}, Y, E and c_0, where $\delta_{\Delta K}$ is the scatter of the test data for ΔK and $\Delta_{\Delta K}$ is the coefficient of variation (COV) representing the uncertainty in the estimated values of the means. The total uncertainty in the fatigue life is then

$$\Omega_z^2 = \Omega_g^2 + \frac{1}{\mu_z^2}\left(c_1^2\sigma_{\Delta K}^2 + c_2^2\sigma_{K_{Ic}}^2 + c_3^2\sigma_E^2 + c_4^2\sigma_{f_b}^2 + c_5^2\sigma_{c_0}^2\right) \qquad (18.4)$$

where $\Omega_g^2 = \delta_g^2 + \Delta_g^2$; δ_g is the scatter of the mean fatigue life of the test specimens, and Δ_g is the uncertainty in the predicted mean life associated with the imperfection in the fatigue theory, including the effects of impact.

18.2 Reliability function

The distribution of fatigue life under a constant stress range follows the Weibull distribution (Ang, 1975). The probability of no fatigue failure or the *reliability function* is then

$$L = \exp\left\{\left(-\frac{N_f}{N_f^*}\right)\Omega\left(1+_z^{1.08}\right)\right\} \qquad (18.5)$$

where N_f^* = the mean fatigue life to be determined theoretically or experimentally, $\Gamma(x)$ is the gamma function, Ω_z = the total uncertainty in the fatigue life, assuming that the minimum performance level is zero. Thus the fatigue is dependent on the mean life and the total uncertainty.

18.3 Typical reliability parameters for fatigue highways

The reliability parameters are: $\Delta K, K_{Ic}, E, Y$ and c_0. The mean and standard deviation of each parameter is discussed briefly.

Cyclic increment in the SIF

ΔK is a function of the increment in the value of the bending stress at the bottom of the pavement and the material properties of the component layers of the pavement system. The modulus of elasticity E and Poisson's ratio υ and thickness H of each layer, are used to determine the stresses as an 18-kip axle load passes over the crack being analyzed by **CHEVRON**. Each has an uncertainty which is considered separately in determining the total uncertainty of ΔK. The uncertainty in the thickness and the Poisson's ratio is not significant when compared to the other variables. Therefore only the mean and standard deviation of Δf_b is taken here, since the mean and standard deviation of E^* is included above. The mean is obtained from CHEVRON and the COV is assumed to be 5%.

Fracture Toughness K_{Ic}

Depending on the magnitude of the *plane strain* fracture toughness, it may not be possible to determine it experimentally in the laboratory. Hence full-scale tests may be necessary. Since there have been no such tests as far as the author is aware, the mean and standard deviation are difficult to assess. Let $\sigma_{K_{Ic}} = 0.15$.

Ultimate bending strength, Y

This is determined by testing beams for the bending strength in the laboratory. Evidently the quality of the mix and the compaction are important. A COV = 0.15 is reasonable.

The size of the starter crack c_0

This parameter is calculated from measurements of the fracture toughness and the bending strength as $c_0 = 0.125 K_{Ic}^2 / Y^2$. As the values of K_{Ic} and Y are already included above, it is only the COV of Ouchterlony equation that is considered here. A value of 15% seems reasonable or $\sigma_{c_0} = 0.15$.

Example 18.1

What is the reliability of fatigue life prediction of the asphalt concrete pavement shown in **Fig. 17.1** The following material properties are given:

Bending modulus E^*	1000 MPa
Fracture toughness K_{Ic}	872 kPa $\sqrt{m}$
Bending strength f_b	385 psi

Solution

The fatigue life is $N_f = \int\limits_{c_0}^{c_f} \dfrac{E^* f_b}{0.153 \Delta K^2}$

where the size of the starter crack $c_0 = 0.125 \left(\dfrac{K_{Ic}}{f_b} \right)^2$ = the failure crack length = 0.65 (h) = 5.2 in. From **CHEVRON** $\Delta\sigma = 28$ psi, and the fatigue life is $8(10)^7$ cycles.

For reliability purposes it is sufficiently accurate to approximate the SIF by

$$\Delta K^2 \approx \frac{\Delta\sigma^2 c\pi}{1 + 0.125\pi^2 c^2}$$

Therefore the fatigue life may be approximated by

$$N_f = \int\limits_{c_0}^{c_f} \frac{E^* f_b}{0.153} \frac{1 + 0.125\pi^2 c^2}{\Delta\sigma^2 c\pi}$$

Taking the COV of $c_0 = 0.15$, gives

$$\frac{\partial N_f}{\partial c_0} = 0.22$$

The other COVs are much simpler

$$\frac{\partial N_f}{\partial \Delta K^2}, \qquad \frac{\partial N_f}{\partial K_{Ic}} = 0.01, \qquad \frac{\partial N_f}{E} = 0.05, \qquad \frac{\partial N_f}{f_b} = 0.05,$$

Therefore

$$\Omega_Z^2 = \left(0.15^2\right) + \left\{\left(0.15^2\right) + 0.05^2\left(0.05^2\right) + 0.05^2\left(0.05^2\right) + 0.22^2\left(0.15^2\right)\right\}$$

So that $\Omega_Z = 0.22$.

$$L = \exp\left\{\left(-\frac{N_f}{N_f^*}\right)\Gamma\left(1+\Omega_Z^{1.08}\right)\right\}$$

$$L = \exp\{-0.22\Gamma(1+0.22\ \}^{1.08} = \exp(-0.22)0.88$$

Therefore the reliability of the fatigue life prediction is $L = 0.82 \leftarrow$ Ans

19

REFERENCES

Allen, D. L. and Deen, R. C. (1986). Computerized analysis of rutting behavior of flexible pavements. *Transportation Record,* No. 1095, 1-10.

American Society of State Highway and Transportation Officials, Interim Guide for Design of Pavement Structures, Washington, D. C, 1972.

Ang, A. H. S. (1974). A comprehensive basis for reliability analysis and design, Japan-USA joint Seminar on Reliability Approach in Structural Engineering, Marzuen Co. Ltd., Tokyo, 29-40.

Ang, H. Folias, E. S. and Williams, M. L. (1963). The bending stress in a cracked plate on an elastic foundation, *Journal of Applied Mechanics,***30**,245-251.

Arthur, J. R. F. and Menzies, B. K. "Inherent anisotropy in a sand". *Geotechnique, 22:* 115 – 129

Arthur, J. R. F., Chua, K. S. and Dunstan, T. (1977). Induced anisotropy in a sand. *Geotechnique, 27:* 13-31

Ballinger, F. (1970). Cumulative Fatigue Damage Characteristics of Plain Concrete. SP 41-1, *Highway Research Board,* Washington, D. C., 48-60.

Barron, R. A. (1948). Consolidation of fine-grained soils by drain wells. *Trans. ASCE,* Vol. 113,1718-1748.

Bazant, Z. P. (1985). Fracture in concrete and reinforced concrete. *Mechanics of Geomaterials, Rocks, Concrete and Soils.* John Wiley, N. Y. 259-303.

Bazant, Z. P. (1996). Analysis of work of fracture for measuring fracture energy of concrete. *ASCE Journal of Engrg. Mech.*, 122(2), 138-144.

Bear, J. (1972). *Dynamics of fluids in a porous media.* Elsevier Science Publishing Co. Inc., New York.

Beukner, H. F. (1970). A novel principle for the computation of stress intensity factors. *Z Agnew Math Mech.*, 59, 529-546

Broek, D. (1986). *Elementary fracture Mechanics*, 4th ed. Martin Nijhoff Publishers, Dordrecht

Broek, D. *Elementary Fracture Mechanics 4th ed.* Martinus, Nijhoff, Boston, 1984.

Brown, S. F. and Snaith, M. S. (1974). The permanent deformation characteristics of a dense bitumen macadam subjected to repeated loading. *Association of Asphalt Paving Technologists*, St. Paul, Minn., 224-252

Budiansky, B., and Hutchinson, J. W. Analysis of closure in fatigue crack growth. ASME Journal of Applied Mechanics, 45, 267-276.

Button, J. W. "Overlay construction and performance using geo textiles". Transportation Research Record, 1248, TRB, Wash. D. C. 1989, 24-33.

Carpenteri, A. (1982). *Engineering Fracture Mechanics*, Vol. 16, No. 4, pp. 467-481.

Chevron Oil Co. Ltd. Ten layer elastic stress distribution in flexible pavements, 1978.

Christian, J. T. (1976). One dimensional consolidation with pore water pressure generation, A Technical Note. *ASCE Journal of the Geotechnical Engineering Division*, **102** (GT 10): 1111-1115

Creed, R. F. (1993). High cycle fatigue of unidirectional Fiberglass composite tested at high frequency. MS thesis, Montana State University.

Croney, D., Coleman, J. D., and Black, W. P. (1958). Movement and distribution of water in relation to highway design. *Highway Research Board, Special Report, No. 40,* 226-252.

Darter, M. (1977). Design of zero-maintenance plain-jointed concrete pavement Vol. I: Development of design procedures. Rep. FHWA-RD-77-III, Federal Highway Administration, Washington, D.C.

Darter, M. Hall, K. T. and Kuo, C. (1995). Support under Portland cement concrete pavements. *NCHRP Transportation Research Board*, Washington, D. C.

DeChang,T. Dan-Quan, L. and Jia-Ju, Z. (1978). Crack propagation under combined stress in three-dimensional medium. *Engineering Fracture Mechanics*, 16 (1), 5-17.

Dorman, G. M., and Metcalf, C. T. (1965). "Design Curves for flexible pavement design". Highway Research Record, No. 71.

Egan, J. A. and Sangrey, D. A. (1978). Critical state model for cyclic load pore pressure. *ASCE Geotechnical Engineering Division, Specialty Conference on Earthquake Engineering,* Pasadena, June, Vol.1, 410-424.

Erdogan, F. and Kebler, K. (1969)Cylindrical and spherical shells with cracks. *Int. J. of Fracture Mechanics.*

Folias, E. S. (1970). On a plate supported on an elastic foundation and containing a finite crack. *Int. Journal of Fracture Mechanics*, 6:3, pp 257-263.

Frederick, D. A. Stress relieving interlayers for bituminous resurfacing. Research Report, NY SDT. Engineering and Research Development, 1984. 37pp.

Gibson, R. E., Schiffman, J. K. L. and Pu, S. L. (1967). Plane strain and axially symmetric consolidation of a clay layer of limited thickness. University of Illinois, MATE Report No. 67-4.

Gorner, F., Mattheck, C., Moraweitz, P. and Munz, D. (1985).). On the limitations of Petroski and Achenbach crack opening displacement approximations of the calculation of weight function. *Engineering Fracture Mechanics* **22,** 269-277

Grandt, A. F. Jr. and Sinclair, G. M. (1971). Stress intensity factors for surface cracks in bending, stress analysis and growth of cracks. *Proc. of 1971 Nat. Sym. on Fracture Mechanics, Pt. 1, ASTP 513,* 1972, pp. 37-58.

Griffith, A. A, (1920). The phenomena of rupture and flow in solids. *Phil. Trans. Royal Soc.,* London A(221), p. 163.

Groenendijk, J., M., Vogelzang, C. H., Miradi, A., Molenaar, A. A. A. and Dohmen, L. J. M. Results of the LINTRACK performance tests on a full-depth asphalt pavement. *Transportation Research Board*, 76th Annual Meeting, Jan, 1997.

Guo, E. H. (2006). Fundamental model of curling response in concrete pavements. 6th Int. Workshop on fundamental modeling of design and performance of concrete pavements, Belgium, Sept. 2006.

Hardin, B. O., and Richart, F. E. Jr. (1963). Elastic wave velocities in granular soils. *ASCE Journal of Soil Mechanics and Foundations Division,* 89(SMI); 33-65.

Heath, A. C., Roessler, J. R. and Harvey, J. T. (2003). Modeling longitudinal, corner and transverse cracking in jointed concrete pavements. *Int. J. of Pavement Engineering*, 4(1), 51-58.

Hetenyi, M. (1946). *Beams on Elastic Foundation.* Ann Arbor, University of Michigan Press.

Hong, A. P., Li, Y. N., Bazant, Z. P. (197). Theory of crack spacing in concrete pavements.

Huang, Y. H. (1993). *Pavement analysis and design.* Prentice-Hall, Englewood Cliffs, N. J.

Hutchinson, J. R. and Paris, P. C. (1973). *Elastic Plastic Fracture.* Stability analysis of J-controlled crack growth. ASTM International, West Conshohocken, PA, 37-61.

Ishibashi, I, Chen, Y. C. and Jenkins, J. T. (1988). Dynamic shear modulus and fabric, part II, stress reversal. *Geotechnique*, **38**, 33-37.

Ishihara, K., Tatsuoka, F., Yasuda, S. and Yasuda, S. (1975). Undrained deformation and liquefaction of sand under cyclic stresses. Soils and Foundations, 15: 29-44.

Journal of Engineering Mechanics, 267-275.

Kelley, E. F. (1939). Application of the results of research to the structural design of concrete pavement. *Public Roads* Vol. 20, No. 6.

Kanninen, M. F. and Atkinson, A. (1980). Application of an inclined strip-yield crack tip model to predict constant amplitude fatigue crack growth, *Int. Journal of Fracture*, 16, 53-69.

Kesler, C. E. (1953). Effect of speed of testing on the flexural fatigue strength of plain concrete. Highway research Board, Vol. 32, 251-258.

Kharlab, V.D. (1990). On a singular strength criterion. *Issledovan Mekhlumike Strol'nykh Konstrukuktrii Mateialov* (Research on Building Structures and Materials), Leningrad, 1990. pp. 82-85. Paper translated from Russian by V, Lazowsky; edited by A. M. Ioannides and I. Khazonovich (private communication).

Lambe, T. W. and Whitman, R. V. (1969). *Soil Mechanics.* John Wiley & Sons, New York.

Li, Y. N. and Bazant, Z. P.(1994). Penetration fracture of floating ice plate: 2D analysis and size effect. *J. Engineering Mech.*, ASCE, 120(7), 1481-1498.

Lin,S. and Folias. E. S. (1975). On fracture of highway pavements. *Int. Journal of Fracture*, **11**, 93-106.

Little, D. N., Lytton, R. L., and Kim, Y. R. Propagation and healing of microcracks in asphalt concrete and their contributions to fatigue. Asphalt Science and Technology, ed. Usmani, A. M. 1997, 527 p.

Liu, S. J. and Lytton, R. L. (1984). Rainfall infiltration, drainage, and load–carrying capacity of pavements. *Transportation Research Record*, No. 993, 28-36.

Majidzadeh K., Bayomy, F. and Khedr, S. (1978). Rutting evaluation of subgrade soils in Ohio. *Transportation Research Record*, No. 671, 75-84.

Majidzadeh K., Buranorom, C. and Karakouzian, M. Application of fracture mechanics for improved design of bituminous concrete, Report FHWA-RD-76-91, 1976, Vols. 1 and II, pp 1-480.

Majidzadeh K., Khedr, S., and Guirguis, H. (1976). Laboratory of a mechanistic rutting model. *Transportation Research Record* No. 616, 34-37.

Morris, J., Haas, R. C. G., Reilly, P., and Hignell, P. (1974). Permanent deformation in asphalt concrete can be predicted. *Assoc. of Asphalt Pavement Technologists.*, St. Paul, 11-76.

Murdoch, J. W. and Kesler, C. E. (1959). Effect of the range of stress on fatigue strength of plain concrete beams. *American Concrete Institute*, Vol.55, 221-231.

Newman, J. C. (1997). Fatigue and fracture mechanics. *ASTMSTP1321D*

Newman, J. C. Jr. (1998). The merging of fatigue and fracture mechanics concepts: a historical perspective. ASTM STP 1321, Fatigue and Fracture Mechanics, 28th vol.

Newman, J. C. and Edwards, P. R. (1988). Short crack growth behavior in aluminum alloy. *AGARD,* 723.

Niu, X. (1990). Some requirements on the reference loading with large stress gradient for the calculation of weight function using the Petroski and Achenbach method. *Engineering Fracture Mechanics,* **36**(1).

Niu, X. and Glinka, G.(1987). On the limitations of Petroski and Achenbach crack opening displacement approximations of the calculation of weight functions-do they really exist? *Engineering Fracture Mechanics,* **26**, 701-726.

Ouchterlony, F. (1983) Fracture toughness testing *Rock Fracture Mechanics*, ed. Rossmanith, New York, pp. 69-150.

Packard (1973). Thickness design of concrete pavements. PCA Bulletin .

Paris, P. and Erdogan, F. (1963). A critical analysis of crack propagation laws. *Journal of Basic Engineering*, 85, pp. 528-534.

Petroski, H. J. and Achenbach, J. D. (1978). Computation of the weight function from a stress intensity factor. *Engineering Fracture Mechanics*, 49 (4), 517—532.

Pickett, G. (1948). A study of the stresses in the corner region of slabs under large corner loads. *Concrete pavement design, App. III.* Portland Cement Assoc., Stokie, ILL

Prevost, J. H. (1978). Plasticity theory for soil stress-strain behavior. *ASCE Journal of Engineering Mechanics Division,* 104 (EM5): 1177-1194.

Ramsamooj, D. V., Correa, A., Sanchez, L., Woodland, J., Kung, J., Chisek, V. Unpublished Project Report. August 2012. " Development and Testing of a new paving material EVAPAVE",

Ramsamooj, D. V, and Piper, R. Theoretical prediction of rutting in flexible pavement subgrades. *Canadian Geotechnical Journal*, Vol.29, 1992.

Ramsamooj, D. V. and Alwash, A. J. (1990). Model prediction of the cyclic response of soils. *ASCE Journal of Geotechnical Engineering,* 116; 1053-1072.

Ramsamooj, D. V. Fatigue cracking of asphalt concrete pavements. ASTM *Journal of Testing and Evaluation, JTEVA,* 1991, Vol. 19, No. 3, May, pp. 231-239.

Ramsamooj, D. V. Fatigue cracking of asphalt concrete pavements. *ASTM Journal of Testing and Evaluation, JTEVA,* 1991, Vol. 19, No. 3, May, pp. 231-239.

Ramsamooj, D. V. Fracture of highway and airport pavements. *Engineering Fracture Mechanics,* 1993, Vol. 44, No. 4, pp. 609-626.

Ramsamooj, D. V. New analytical method for prediction of fatigue cracking of asphalt concrete. *International Journal of Pavement Design,* 2000.

Ramsamooj, D. V. (1993). Fatigue cracking and rutting in flexible pavements. *ASCE Conference on Fracture Mechanics Applied to Geotechnical Engineering,* 132-146.

Rice, J. R. and Johnson, M. A. (1970). The role of large crack tip geometry changes in plane strain fracture. *Inelastic Behavior of Solids,* Kanninen et al. ed. 641-672.

Richart, F. E. (1959). Review of the theories of sandrains. *Trans. ASCE*, Vol.124., 709-736.

Ridgeway, H. H. (1976). Infiltration of water through the pavement surface. *Transportation Research Board*, 616,98-100.

Roscoe, K. H. and Burland, J. B. (1968). On generalized stress/strain behavior of wet clay. *Engineering Plasticity* ed. Heyman, J. and Leckis, F. A. Cambridge University Press, U. K., 535-600.

Rowe, P. W. (1971). The stress-dilatancy relation for static equilibrium of an assembly of particles in contact. *Proc. of Royal Soc., Series A*, London, England, 269, 500-527.

Sangrey, D. A., (1968). The behavior of soils subjected to repeated loading. Ph.D. thesis, Dept. of Civil Engineering, Cornell University, Ithaca, N. Y.

Sangrey, D. A., Henkel, D. J. and Esrig, M. I. (1969). The effective stress response of a saturated clay to repeated loading. Canadian Geotechnical Journal, **6**: 241-252.

Schapery, R. A. (1975a). "A Theory of Crack Initiation and Growth in Viscoelastic Media I Theoretical Development, *International Journal of Fracture*, 11, pp. 141-159

Schapery, R. A. (1975b). A Theory of Crack Initiation and Growth in Viscoelastic Media II Approximate Methods of Analysis, *International Journal of Fracture*, 11, pp. 369-388.

Schapery, R. A. (1975c) A theory of crack initiation and growth in viscoelastic media iii analysis of continuous growth, *International Journal of Fracture*, 11, pp. 549-562.

Schapery, R. A. (1984). Correspondence principle and J-Integral for large deformation and fracture analysis of viscoelastic media, *International Journal of Fracture*, 25, 195-223.

Seed, H. B. 1976 Evaluation of liquefaction effects on level ground during earthquakes. *ASCE Annual Convention and Exposition on Liquefaction,* Philadelphia, 1-104.

Sih, G. C. and Liebowitz, H. (1968). Mathematical theories of brittle fracture. *Fracture-An advanced treatise*, Vol. II. Academic, New York, 1-190.

Sih, G. C. and Setzer, D. E. (1964). Discussion of the paper on "The bending stress in a cracked plate on an elastic foundation" *J. Appl. Math*, 30, 365-367.

Taylor, D. (1989. *Fatigue Thresholds*. Butterworths, London, p.218

Terzaghi, K. (1943). *Theoretical soil mechanics*. John Wiley, N. Y.

Terzaghi, K. (1955). Evaluation of coefficients of subgrade reaction. *Geotechnique*, 5, 297-326.

Tseng, K. and Lytton, R.L. (1990). Fatigue Damage Properties of Asphaltic Concrete Pavements. Presented at the 69th *Annual Meeting of the Transportation Research Board,* Jan. 1990 43p.

Tseng, K., (1988). A finite element method for the performance analysis of flexible pavements. Ph.D. Dissertation. Texas A&M University, College Station.

Westergaard, H. M. (1925). Computations for stresses in concrete roads. *Highway research Board*, 90-112.

Westergaard, H. M. (1948). New formulas for stresses in concrete pavements. *ASCE Trans.* Vol. 113.

Wnuk, M. P. (1971). Subcritical growth of fracture. *International Journal of Fracture Mechanics,* Vol. 4, pp 383-407.

Wnuk, M. P. Prior-to-failure extension of flaws under monotonic and pulsating loadings. *Engineering Fracture Mechanics*, 1973, Vol. 5, No. 5, pp. 379-396.

Wolfram, S. (1993). *MATHEMATICA*. A system for doing mathematics by computer, 2nd Edition.

Wu, E. M. A fracture criterion for orthotropic plates under the influence of compression and shear. TAM Report 238. University of Illinois, 1965.

Yoder, E.J. and Witczak. M.W.,1975. *Principles of Pavement Design*, John Wiley and Sons, Inc.

Zeevaert, L. (1976). *Foundation engineering for difficult subsoil conditions*. Van Nostrand Reinhold Co. Inc., New York.

20

NOTATION

The following symbols are used in this book:

A	pore water pressure parameter
A^m	Skempton pore water pressure coefficient for the mth yield surface
A^e	elastic range of A
a, c	semi-elliptical minor and major axis of crack
c_0	size of the inherent defect or starter crack
C_{ce}^1, C_{se}^1	elastic compressibility in the axial and expansibilty in the lateral direction, resp.
d	maximum size of the aggregate
D	flexural rigidity
e	void ratio
E_b^*, E_c^*	dynamic modulus in bending and cylinder in compression, respectively
f	yield function
f_{ct}	indirect tensile (cylinder splitting) strength
f_b	bending stress in the uncracked beam
G, Gc	strain energy release rate, and its critical value
G_s	specific gravity g plastic potential
H	plastic modulus
H'	plastic shear modulus
Hn	thickness of the nth layer
H_{dr}	length of the longest drainage path for consolidation
k	coefficient of subgrade reaction

k^m, k^p	size of the *m*th and outermost yield surfaces, respectively
K_I, K_{Ic}	stress-intensity-factor in the opening mode and its critical value
K_0	threshold stress intensity factor corresponding to the endurance limit
R	ratio of minimum and maximum SIFs
p	effective mean pressure
P	vector normal to the yield surface
q	deviator stress
Q	vector normal to the potential surface
s	soil suction
s	deviator stress tensor
T, T'	time-factor for 1-D and 3-D consolidation, respectively
u, Δu	pore water pressure and its increment
w	deflection of the concrete slab
$3/2\alpha_z, \beta$	centers of the yield surfaces in the deviatoric plane and the hydrostatic axis
δ	Kharlab's characteristic length for materials
ε	total strain
$\varepsilon_v^e, \ \varepsilon_v^p, \varepsilon_v$	elastic, plastic and total volumetric strain
$\varepsilon_d^e, \varepsilon_d^p, \varepsilon_d$	elastic, plastic and total deviatoric strain
$\in (t)$	creep compliance function
υ	Poisson's ratio
w	deflection of the concrete or AC slab
λ	inverse characteristic stiffness
κ	rebound compression
γ_w	unit weight of water
η	elastic portion of the ratio of the compressibility and expansibility
ρ	radial distance from corner of the slab
θ	first invariant of the stress tensor
σ, τ	normal and shear stress
ψ	H_{dr}^2 / c_v times the rate of generation of the excess pore water pressure

CONVERSION FACTORS FOR U.S. UNITS AND SI UNITS

Quantity	U.S.	SI units
Area	ft^2	$0.0929\,m^2$
Energy	ft.lb	1.356 J
Force	kip	4.448 N
Length	ft	0.3048 m
Mass	slug	14.59 kg
Moment of force	lb.ft	1.356 N.m
Stress	psf	47.88 Pa
Volume	ft^3	$0.02832\,m^3$

COMPUTER PROGRAMS

D. V. RAMSAMOOJ WEB SITE

This book contains nine (9) computer programs that are an integral part of the book

All computer Programs and only Computer Programs are typed in *BOLD ITALICS*. Each computer program has the input data for the solution of a particular program in U. S. units. The programs are all editable, so that other problems can be solved by inputting the proper numbers.

I   *CHEVRON* Multilayered Stress Analysis. Computes the stresses, strains and deflection in a multilayered pavement. The width of the outer lane (surface and base) should be continued for a distance of 7 times the thickness of the surface layer, to eliminate the edge effect.

II. *WESTRESS* Westergaard Stress Analysis for Rigid Pavements. Computes Westergaard stresses for rigid pavements.

III.*EFM* Elasticity and Fracture Mechanics. Computes the relationship between cracks and the change in the deflection of the pavement.

IV. *DEFCRACK* Computes the change in the deflection caused by cracking in component layers.

V. *ALLIGATOR* Flexible Pavement Fatigue. Determines the fatigue life of flexible pavements (Ramsamooj et al, 2012)

VI. *SUBGRADE* Rutting in Pavement Subgrades. Computes the permanent deformation in subgrades.

VII. *RUT.* Rutting in Flexible Pavement Layers. Computes the permanent deformation in AC or other flexible pavement layers

VIII. *CONPAVE.* Rigid Pavement Fatigue. Determines the fatigue life of concrete pavements

IX. *ALLIGATOReva.* Design of highways and airports using EVAPAVE.

These computer programs are stored in a website named www.dindialramsamooj.net Password:dindial495237.

ACKNOWLEDGEMENTS

The author is deeply grateful to President Milton Gordon of the California State University, Fullerton, for providing research funds for developing the new material EVAPAVE. He is also grateful to Dr. Unnikrishnan, Dean and to Dr. P. Rao, C.E. Chair, for support in the laboratory research. The fine work of Adrian Correa, graduate student, Leroy Sanchez, Jimmy Kung, and Vadim Chisek, Directors of Civil Engineering, Electrical Engineering, and Computer Science Laboratories, respectively, are gratefully appreciated. Ms. Kelly Donavon, graphic artist, did all of the fine graphic illustrations.

INDEX